T0356485

Project Workforce Estimating

Project Workforce Estimating

Best Practices for Project Managers

Harold Kerzner, Ph.D. and
Al Zeitoun, Ph.D.

Published by John Wiley & Sons, Inc., Hoboken, New Jersey.

Published simultaneously in Canada.

Library of Congress Cataloging-in-Publication Data applied for:

Hardback ISBN: 9781394319374

Cover Design: Wiley
Cover Image: © Lyndon Stratford/Getty Images

SKY10099102_022825

To
Our wives – Jo Ellyn and Nicola
And our kids –
Jason B, Lindsey, Andrea, Jacalyn, Jason K,
Adam, Zeyad, and Sarah
For being our source of joy and continued source of inspiration

Contents

Preface *xi*
About the Companion Website *xiii*

1 The Future of Project Workforce Planning *1*
Navigating the World of Limited Resources *1*
Principles of Project Workforce Management *3*
Workforce Management in the External Environment *6*
Workforce Management and Legislation *7*
Professional Development for Project Teams *8*
Labor Rate Structures *9*
The Role of Contract and Temporary Staff *10*
Impact of Artificial Intelligence on Future Workforce Planning *11*

2 The Complexities of Project Workforce Estimating *17*
Sources for Workforce Estimation *17*
Factors Influencing Workforce Estimation *18*
Stakeholder Involvement in Workforce Staffing *20*
The Key to Workforce Estimating *21*
Earned Value Management Systems and the PMBOK® Guide *22*
Direct Versus Indirect Project Costs *23*
Breaking Down the Overhead Costs *24*
Forward Pricing Rates: Salary *25*
Calculating Available Work Hours *26*
Work Authorization Form *27*
Project Pricing Overview *27*
Validating Estimation Assumptions *28*
The Fuzzy Front End *30*
Project Portfolio Management *32*
The Value Proposition Behind Project Portfolio Software Tools *33*

3 Techniques for Estimating Project Workforce Needs *35*
Overview of Workforce Estimation Methods *35*
From Labor Hours to Labor Costs *36*
Enhancing Estimation Accuracy *36*
Estimating Costs Per Hour *37*
Parametric Estimating *38*
Analogy Estimating *38*
Ground-up (Grassroots) Estimation *39*
Applying Learning Curves in Workforce Estimation *40*
Understanding the Learning Curve Effect *41*
Estimating Management and Support Needs *42*
Identifying Hidden Labor Costs *43*
The Impact of Documentation on Labor Costs *44*
The Need for Workforce Backup Plans *45*
Common Challenges in Workforce Estimation *46*
The Essential Value of Enterprise Risk Management *47*

4 Monitoring Workforce Expenditures *49*
Initiating Workforce Expenditure Tracking *49*
Converting Work Hours into Financial Metrics *50*
Balancing Hours and Dollars in Project Tracking *50*
Analyzing Workforce Metrics *52*
Analyzing Spending Trends *53*
Example of Termination Liability *55*
Oversight of Workforce Expenditures *55*
Setting Reporting Intervals for Workforce Status *57*
Documenting Challenges in Workforce Reporting *57*
Optimizing the Business Models Around Workforce Strengths *58*

5 Growth of Innovation Project Teams *61*
The Need for Innovation and Creativity *61*
Introduction to Innovation *62*
Types of Innovation *64*
Co-Creation Innovation *69*
Defining Innovation Success and Failure *72*
Value: The Missing Link *74*
The Innovation Environment *74*
The Innovation Culture *77*
Idea Generation *78*
Understanding Reward Systems *79*
Resources Management *80*

The Power of the Agile and Hybrid Approaches 83
Innovation Project Management Future Skills 84
Innovation Portfolio Management 87
The Need for Innovation Metrics 89
Extracting the Business Value 93
The Value of Prompt Engineering in Fostering Innovation 95
References 96

6 Designing the Future Workforce 101
Identifying Desired Team Competencies 101
Perspectives on Project Management 102
Strategies for Workforce Recruitment 103
Degrees of Permissiveness 103
Commitment and Expectation Management 104
Engaging High-Value Team Members 105
Addressing Underperformance 105
Non-Financial Incentives for Motivation 106
Celebrating Achievements from Plaques to Public Acknowledgments 106
The Role of Public Recognition 107
Alternative Non-Monetary Rewards 107
Timing and Patterns of Staffing Needs 109
The Role of Organizational Development 110

7 Advanced Topics in Workforce Planning 115
Budget Allocation and Adjustment 115
Securing Additional Project Funding 116
Global Workforce Estimation Challenges 116
Structuring Project Teams for Success 117
Incorporating Management Plan Data 118
Developing Contingency Plans for Workforce Management 119
The Benefits of Co-Located Teams 120
Project Life Cycle Costing Approaches 121
Techniques for Workforce Leveling 121
Advanced Workforce Leveling Strategies 122
Maturing an Inventory of Skills and Competencies 123
The Next Potential of AI in Enabling Workforce Planning 125

8 Case Studies in Workforce Planning 133
Dixon Aerospace 135
The Phoenix Project 137
Brenda's Dilemma 139

The Brainstorming Meeting *141*
The Lack of Information *143*
The Information Overflow Dilemma *145*
The Impact of Assumptions *147*
Nora's Dilemma *149*
Managing Resources in Government Agencies *153*
Kane Corporation *156*
Skunk Works Project Management *160*

Index *173*

Preface

A great deal of literature exists on workforce planning and most of the analyses contain models for matching the supply and demand for labor as organizations attempt to grow. The models also discuss how costs and productivity will be impacted.

Most of the models address a production environment where the demand is known or can be predicted with reasonable accuracy. The costs are attributed to hiring workers that require training to achieve the skills needed. Training can take place on-the-job or off-the-job. The models may also discuss labor shortfalls resulting from random resignations and when employees retire.

In a project environment, workforce demands are highly uncertain. There exists a shortage of literature on workforce planning in project environments. There is no guarantee in a project environment that clients will ask for other similar products or services once your project for them is completed. Simply stated, project environments usually have an ever-changing demand for products and services. New clients may emerge that require a workforce with different skills. We must either retrain the existing workforce or terminate some of them and hire new recruits to be trained.

The type of project also impacts workforce planning. Projects for external clients may be the result of competitive bidding with the goal of achieving a certain level of profitability. These projects are needed for strategic organizational growth. At the same time, there may be internal or operational projects that must be staffed for the business to continue. Continuous competition for resources between internal and external projects can occur.

Project workforce teams of the future will have to accommodate the impact of digitalization. Technology enhancements and augmentation are here to stay, and the future project workforce has a wonderful opportunity to make technology a true ally. This has the potential of making the workplaces of the future more fun. Instead of the classic nervousness of project teams about the availability or accuracy of the data, or if the boos will like the report, we could shift the focus to other critical discussions and to building future skills, like creative thinking. In an era when topics like mental health have dominated discussions, it would be great that we design the right fitting workforce mix and that we turn possible project teams' despair to strategic focus and clarity.

The intent of this book is to focus upon workforce planning in a project environment where significant fluctuations can occur in demand, risks, and the business environment. The urgency surrounding this topic stems from the highly projectized future of work that continues to accelerate. This is also coupled with the valuable ongoing exploration around the impact of digitalization on the future jobs to be done and their implications on the future workforce.

About the Companion Website

This book is accompanied by a companion website:

www.wiley.com/go/Kerzner_ProjWFE

The website includes:

- Learning Objectives and figures for each chapter.
- Instructor's Manual: Answers to case studies and teaching notes.

1

The Future of Project Workforce Planning

LEARNING OBJECTIVES
• Understand the challenges with a limited workforce • Understand where the workforce comes from • Understand how legislation impacts the workforce • Understand the need for workforce professional development and future shifts

Keywords *Artificial intelligence (AI); Contracted workers; Corporate workforce needs; Labor rates; Resources limitations; Workforce gap analysis; Workforce legislation; Workforce planning models*

Navigating the World of Limited Resources

Today we live in a world of limited resources. Companies are running lean and mean due to uncertain economic conditions, unavailable qualified labor, and rapidly changing customer demands. Yet companies never seem to run out of projects to work on, but they do have a shortage of resources to support all of the desired projects.

Newly appointed project managers often willingly accept project management positions with the mistaken belief that they will have all of the necessary resources for their projects. Executives and sponsors also reiterate these words, namely that you will get all of the workforce support needed when selecting and appointing the new project manager. But then, after a go-ahead, reality sets in and the project manager discovers that he/she is living in a world of limited resources.

Project Workforce Estimating: Best Practices for Project Managers, First Edition.
Harold Kerzner and Al Zeitoun.

Companion website: www.wiley.com/go/Kerzner_ProjWFE

To make matters worse, newly appointed project managers do not seem to have any idea as to the complexities with project resources staffing and estimating. Wanting an army of resources may seem like a good idea at first, but the allocated budget may not even allow you to have the minimum workforce you think you need. In an ideal situation, you would determine the workforce needed first, and then price out the workforce to determine the budget for the project. While this sometimes happens, it is more likely that the budget is established first by senior management when approving the project, often without any involvement by the project manager, and then the project manager must staff the project based upon the available funding. The result is often a project team with a shortage of resources or team members with inadequate skill sets.

For simplicity's sake, companies can be classified as project-driven and non-project-driven. Project-driven companies usually survive on the various projects they manage for external clients through a competitive bidding process. In these companies, the size and type of resources can fluctuate based upon the types and quantities of projects they are asked to manage.

In non-project-driven companies, there are usually standard production lines, and projects exist to support the creation of new products or modifications to existing products as well as ongoing business needs. Workforce management is somewhat easier in this type of company.

Both types of companies must deal with the risks of limited resources and need to adopt a workforce planning model. There are two reasonable solutions expected from workforce planning based upon limited resources:

- Make sure that we assign the right people with the right skills to the right tasks
- Try to increase productivity, efficiency, and effectiveness

With limited resources, it is essential that we have the right people assigned to the right tasks. Project managers may not know the capabilities of the assigned workers and may have to rely upon the expertise of the functional managers who provide the staff. Increasing the productivity of the assigned workers does not mean producing more deliverables or increasing production. Instead, it implies getting workers to perform their assigned tasks more efficiently or more effectively. The proper investment in training and education can make this happen.

For companies that survive on competitive bidding, limited resources are almost always a way of life. Companies tend to bid on more jobs than their resources can support because they know that they will not be awarded all of the contracts they are bidding on. If they were to win more projects through competitive bidding than they can handle, there would still be a reluctance to hire more people for fear that there would be no place to put the people after the projects are completed. Companies that hire when they win a contract and then lay off the workers when the contract is finished may find it difficult to attract talented workers who want some degree of employment stability and security. This can be devastating to the company's reputation and create havoc with workforce planning.

Principles of Project Workforce Management

Workforce management begins with workforce planning. Workforce planning, also known as human resource planning or manpower planning, is the process of determining the human resources that an organization needs to meet its strategic goals. The three critical elements in the process are the forecasting of future labor demand, analyzing present labor availability, and effectively managing resource supply versus demand. The outcome, if done effectively, should be a plan that ensures that the right people with the right skills are assigned to the projects such that there is a high expectancy of achieving the organization's strategic goals. Effective manpower planning also minimizes the risks of overstaffing, having to pay for excess staff that may not be needed, and loss of productivity.

A simple model for future workforce planning is shown in Exhibit 1.1.

Most of the principles in Exhibit 1.1 apply to workforce planning in any type of company.

The focus of this book is future workforce planning for projects.

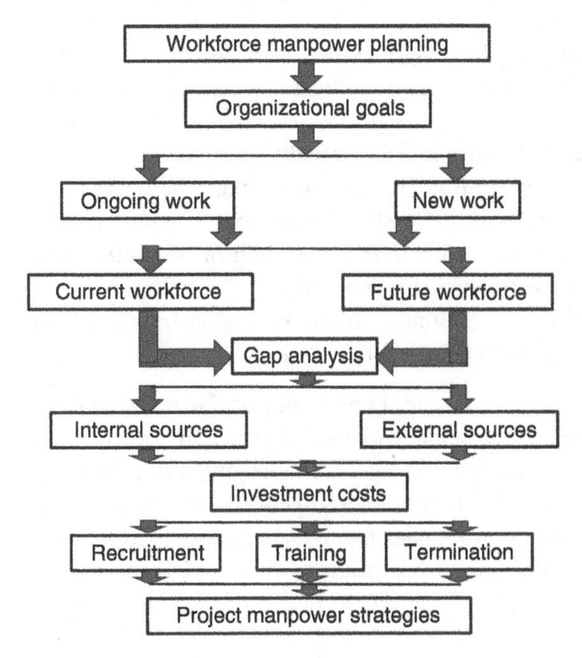

Exhibit 1.1 Future Workforce Planning Model

Workforce planning begins with an understanding of the organization's business goals now and possibly in the future. Forecasting future manpower needs requires answering the following questions:

- What ongoing or new types of projects will be worked on in the future?
- How many people will be required to meet the needs of present and future projects?
- What skill sets will the employees need?

Gap analysis, as identified in Exhibit 1.1, is more than identifying a potential shortage of resources. It also includes answering the following questions:

- Will the organization be required to work on new types of projects?
- Will new skills be required?
- Will training be a necessity for existing or newly hired personnel to develop the skills needed?
- How long might it take for employees to develop the new skills?

Gap analysis is more than making sure you have the right number of resources. It also provides guidance for making sure you have the best possible employees assigned to the best possible activities.

Developing either organizational or project manpower strategies includes the following[1]:

- **Organizational restructuring:** this may include organizational redesign to fit a potentially new business model, regrouping activities, and improving efficiency.
- **Training and development:** this may include providing the current staff with training and development opportunities to encompass their new roles and responsibilities as well as training newly hired workers in the new skills needed.
- **Recruitment activities:** this may include the recruiting of new hires who already have the skills or are willing to learn new skills.
- **Outsourcing activities:** this may include teaming with other individuals or organizations that possess the needed skills for the tasks.

People come and go for a variety of reasons. Long-term workforce planning for projects must consider the risks of replacing workers if needed and how they will be trained. People usually find working on projects challenging and rewarding. This is especially a critical attraction point for the next-generation workforce that enjoys exploring new outcomes. Proper investment in human resources helps retain talented workers and reduces employee turnover.

1 Adapted from Wikipedia, The Free Encyclopedia: strategic human resource planning.

Human resource planning must also include ways to retain talented employees and keep them motivated. Some techniques include:

- Engaging employees effectively when they are assigned to a new project
- Making sure they are assigned to challenging work
- Providing a recognition program for excellent performance
- Providing training that is aligned with their career goals

If we look at the principles of human resources management and project objectives together, we can define project staffing as the process that ensures that the organization has the correct number of people with the correct skills doing the right activities at the right time to achieve the project's objectives. But human resources management on a project is more than simply getting people assigned to the project team. It also involves:

- **Compensation:** Even though project managers may not have any responsibility for wage and salary administration, they might still provide rewards for the team members, whether they are tangible or intangible, and monetary or nonmonetary. They may also be asked to provide informal feedback to the functional managers as to how the workers are performing.
- **Safety and health:** These are things that the project manager must do to protect the employees from on-the-job injuries and work-related illnesses. This also involves providing the workers with a suitable place to work assuming that they are removed from their functional area as would be the case with a co-located team.
- **Training and development:** Companies must recognize the return on investment in employee education and provide people the opportunity to attend classes or other educational opportunities related to improving productivity. This also involves allowing the workers to take time away from the project for education related to their career development opportunities. If the costs of training are directly related to the project, then the project may incur the training costs. If the employees are attending college classes, such as for graduate degrees, then the project manager must allow these people to continue attending classes. This may require adjusting the times when the worker will be performing his/her project tasks.
- **Labor relations:** Project managers should not violate corporate labor relations policies established by their company. This involves fairness in treating employees, discipline, promotions, layoffs, and termination. If the company is unionized, then there are corporate expectations on how the project manager should interact with employees of the union. For example, the union may not want their members assigned to tasks above what their pay grade indicates, especially if the workers are willing to accept the assignment believing they will be immediately promoted.

There are many types of workforce planning models like Exhibit 1.1. Examples include:

- Manpower planning at the corporate level
- Manpower planning at the division or section level
- Manpower planning at the project level
- Short-term manpower planning
- Medium-term manpower planning
- Long-term manpower planning

Each type of plan may have different requirements, different personnel, and constantly changing goals and objectives.

Workforce planning models, as shown in Exhibit 1.1, have been used quite extensively for functional organization human resource planning. Only recently, has the model been applied to project organizations.

In the past, most projects relied heavily upon functional organizations to provide the necessary resources and, when the project was completed, the resources would return to their respective functional units. Today, many of our projects are longer in duration and many projects are treated as temporary functional units requiring complex workforce modeling. As such, models as shown in Exhibit 1.1 are being modified for applicability to project workforce planning, regardless of the size and length of the project.

Workforce Management in the External Environment

On short-term projects, the project manager relies upon the line managers, project sponsors, and the corporate human resources organization to worry about how the external environment influences staffing. But on long-term projects, especially those that require the use of contracted resources during the project, the project manager must be aware of the external environment. This is extremely important if the project is being executed in another country. Some examples include:

- **Economic health:** This includes the economic conditions in the host country as well as economic conditions in the parent country. During favorable economic times, quality resources, which may be limited, are in high demand. In some emerging markets, people may change companies quickly, without notice, and with little regard for the project. Also, in favorable times, companies work on more projects making the best resources in high demand throughout the company. During unfavorable conditions, there is a larger than normal risk that the project might be canceled.

- **The labor market:** During favorable economic conditions, people in the labor market will be seeking higher pay than perhaps what you budgeted for in your proposal. Also, a senior engineer in one country may not have the same skills as a senior engineer in another country.
- **Competitors:** If competitors monopolize the market where you are executing the project, you may be unable to obtain qualified resources. You also stand the risk of losing qualified resources to the competition.
- **Technology:** Technological advances as well as organizational process assets may be limited in the host country. If you have contracted labor, they may not be knowledgeable with these processes.
- **Unions:** Both internal and external unions have tremendous power. Seniority is important and the union may dictate who will work on your project. The union can limit productivity increases and remove people off of your project at the most inopportune time. The union can also prevent employees from working overtime.
- **Society and politics:** Politics and culture in projects for a host country other than yours may create problems. The decision-making process may be quite slow, and staffing may be based upon membership in the right political party. The local government may require that procurement contracts be given to companies within the host country just to keep people employed even though more qualified resources exist elsewhere.

Workforce Management and Legislation

Corporate human resources personnel and most functional managers either understand staffing legislation or are trained in it. Project managers, on the other hand, are often placed into project management roles with little knowledge of staffing legislation, and the results can create serious problems for the parent company. The situation becomes more complicated if contracted labor is used, especially from countries in which the project manager has limited knowledge of the culture and the laws.

For example, in some countries, the workers have a right to hold a job and do not believe that they can be fired even if their performance is subpar on the project. Project managers may not be able to have these people removed from the project. In the United States, as well as in other countries, there have been several laws enacted over the past several decades that can affect staffing practices. Many of the laws discuss hiring and firing practices, discrimination, worker rights, and the use of overtime. It is essential for the project managers to develop a proper degree of awareness, and as needed, supplement their understanding gaps with the right experts dependent on the specific global setting of the project.

TIP With the world of work becoming more global, invest in yourself and expand your view of the staffing practices and the dynamics of the global marketplace.

Professional Development for Project Teams

In the early years of project management, worker career development was the responsibility of the worker or the worker's supervisor. Project managers viewed the workers as if they were contracted labor that was leased from the functional departments and treated as merely a cost to the project to be removed as soon as possible. Today, project managers are expected to help the team improve their performance as well as assist the team members with personal career development opportunities.

Today, project managers are expected to:

- Identify workers that have the potential to improve through additional training
- Provide workers with time away from the job to attend training classes
- Identify workers whose performance warrants promotion or assignments with added responsibility
- Identify workers whose performance warrants a demotion, lateral transfer, a position of lesser responsibility, or even termination
- Determine the need for disciplinary action
- Provide the functional managers with performance appraisal information on how the workers are performing on the job

Project managers may not possess formal reward or penalty power. Project managers may be able to identify each of the above items, but the functional managers may be the only people fully authorized to make the final decisions. The best that the project manager can do might be to provide recommendations to the functional managers.

In the future, this could be one of those major transformations in the role of project managers. With the continual move toward projectized organizations, there is a direct impact on how the design of the future organization will be done. This is coupled with the impact of Artificial Intelligence (AI) and other aspects of digitalization that could directly affect the need for classical business functions and result in more project focus in the design. This would be a turning moment in what is expected of the project manager in the future in relation to the project workforce planning aspects.

TIP The future role of the project manager in workforce development is changing. In a digitally-driven workplace, project managers are able to make better strategic choices.

Figure 1.1 The Future of Professional Development

In addition to the role of driving professional development, recommendations by the project manager in this potential future transformation, as depicted in Figure 1.1, are changing. There is a highly changing nature to professional development. This is driven by the intersection of the real and digital worlds at the center of it. Future professional development will also shift to a strong focus on creativity and innovation. This is stemming from the dominating projectized way of working. Project managers will be the most fitting leaders to recommend what capabilities are critical to nourish for the future workforce.

Labor Rate Structures

The labor rate structure of the company is critical when staffing projects and estimating the associated manpower costs. The greater the number of pay grades in each functional area, the greater the complexity in determining the labor costs, especially if several pay grades can perform the required activity. Although most companies have four or five pay grades per functional area, some companies may have up to ten pay grades with job descriptions that may or may not overlap. As an example, consider an engineering department with the following pay grades:

- Engineering apprentice
- Engineering aide
- Engineering mathematician/statistician
- Engineering technician
- Junior engineer
- Engineer

- Research engineer
- Senior engineer
- Engineering consultant

Assuming these pay grades are listed in ascending order of importance, the difference in yearly salary between the top and bottom pay grades could be more than $100,000 and yet several pay grades may be qualified to perform a given task.

In a digitally enabled future, this is getting even more complicated. Some of the classical roles that will be easily handled by AI might have a negative or a positive impact on labor rates depending on the types of redesigned future jobs to be done. The increasing shift toward focus on value in projects also adds an important dimension to what will be prioritized in considering adjustments to pay grades and the associated labor rates.

The Role of Contract and Temporary Staff

Companies often find themselves in positions where they have a shortage of resources and, instead of hiring permanent employees, they hire temporary workers. This occurs when:

- A company has a need for a specialized skill for a few months and the company does not believe that they will need this skill in the future.
- A company has a need for a specialized skill but only sporadically over the next few years. The cost of hiring someone full-time with this skill is not cost-effective.
- A company has a common activity that requires several engineers. If the company has a shortage of five engineers, then they may hire five temporary engineers to perform this common task.

Temporary workers are hired under a contract that guarantees them employment for a period, usually not less than three months. The temporary workers may find it necessary to pay their own health care costs, taxes, and retirement costs. The contracts usually have a clause that allows for extension of the contract if both parties agree.

Full-time employees generally dislike temporary workers because they soon discover that the temporary worker may be earning a higher base salary than they are. As an example, consider a worker who is earning $60/hour and discovers that the temporary worker is earning $80/hour. The permanent worker may be upset at this difference because he/she did not consider the fully loaded cost.

If the permanent worker has an overhead rate of 150%, then the permanent worker is fully loaded at $150/hour, which includes medical benefits and retirement plan contributions. The temporary employee, on the other hand, must pay for his/her own medical and retirement plans and may have to pay his/her own relocation expenses as well. The temporary employee may not be paid for sick days taken. The temporary employee also risks the chance of unemployment when the existing contract expires.

Project work, in the expanding gig economy, by its temporary nature, opens the door to this type of contract work. In addition to dealing with organizational political aspects, such as the example above, this provides an upside of injecting new areas of expertise into the workforce. Some of the objectivity brought in by those workers could have a direct impact on the bottom line and could be factored into future project estimating.

TIP The impact on culture and culture fit are two valuable aspects of the gig economy. Contractors with diverse expertise and views enrich the culture of the organization.

Another valuable aspect of the contract workers is providing an option for testing cultural fit that would safeguard the higher opportunity of success for these workers if they turn into part of the permanent staffing of the organization. With more people opting for contract work, there have been great development in the platforms that simplify this contract work. Both workers and organizations are gaining the flexibility and freedom highly valued by the future workforce.

Impact of Artificial Intelligence on Future Workforce Planning

As we focus on improving the planning for the future workforce and to tackle many of the challenges and potential risks highlighted above, it would be beneficial to look at project workforce planning from an organizational development angle. Organizational development could be the vehicle that creates the effectiveness roadmap organizations need to be more productive in the future and become more predictable in assessing the fluid future demands on the workforce. There are multiple areas to organizational development that should be looked at and integrated to provide the utmost value to enhancing workforce planning and utilization. This development is now coupled with the inclusion of the impact AI could be bringing into the equation.

To explore some of these areas that directly contribute to the success of workforce planning, let us focus on assessments, leadership, change management, learning, team dynamics, and driving toward excellence.

As we tackle assessing the organization, we recognize that data is the key to leading in the future. Conducting proper diagnosis of the current state of the organization is key to understanding the various strengths, weaknesses, opportunities, and threats. As shown in Figure 1.2, there is a need to have as much objectivity in the

Figure 1.2 Diagnosing the Organization

assessment as possible. This could be done using surveys, polls, interviews, and as many ways as needed to get a clear and holistic view of how ready the organization is to support staffing decisions for projects in the various types of organizations. Culture is a crucial component of this assessment.

Understanding the culture allows for a better view of the true needs of the workforce. This increases the chances of finding the right fitting resources. This is not merely focused on skillsets and capabilities. This has much to do with personality, attitude, and driving values. Proper diagnosis could get into multiple areas of details that would enhance the likelihood that the resource estimates are meaningful. The diagnosis could also illustrate the organizational ability to use AI. Whether AI is in the mix to increase efficiencies, speed of accomplishing long-duration tasks, or conducting routine reviews, this enhancement has a direct impact on how workforce demands look like. AI could also be an instrument for speeding up research and development activities and thus contributing to shortening product lifecycles and directly affecting the type and amount of future workforce needed.

The potential outcomes of properly assessing the organization could be:

- Enhancing the chances for proper workforce fit
- Designing the future organization in the most aligned way with the project needs
- Increasing the likelihood of project workforce estimating accuracy
- Improving the motivation, morale, and productivity of the workforce
- Clear understanding of readiness for the use of AI

Leadership is also drastically changing in the future organization design. Leading in the digital era is not the same as its predecessor models. Not only are the leaders expected to have digital fluency or at least comprehension, they are expected to put digital in practice to create higher value to the organization's shareholders and stakeholders. Leading in this digital era also requires a higher degree of humbleness reflected in the lifelong learning style that the leaders should exemplify across the future workforce.

This new focus on the diverse human leadership qualities of the superhero project managers in the digital age will continue to increase. Figure 1.3 shows a set of those qualities that continue to be valued in the leadership models expected in the future organization. As we try to better predict the future workforce demands, it is essential for us to evaluate these key ingredients in the figure such as empathy, curiosity, and gratefulness. These ingredients have a direct correlation to how productive the workforce is and how this can ultimately affect the accuracy of the estimating process.

Managing change is another dimension that is dominating the organizational design and the use of AI in the future. Projects and Programs constantly create change. The skills and qualities necessary to manage change and adapt to the changing working environment and environmental factors surrounding project work, put a high demand on considering that in future organization design and selection of workforce. AI could help us with patterns and generative meaningful data that complement staffing decision-making and the ideal workforce mix of the organization. Every attempt should be made that we capitalize on data to educate the estimating process from lessons learned to continual enhancements in productivity possibilities.

Figure 1.3 Future Leadership Qualities

TIP The design of the future organization that is highly digital will result in enhancing the selection and estimating of the future workforce.

Learning as a cornerstone of the future organization will empower the project estimating of the future workforce. This is an area where one could tremendously expedite and improve the planning process. Using data and patterns from across different sizes and complexities of projects to more accurately estimate is a difference maker in the future organization. Leaning organizations have unique cultures that naturally require different sets of workforces. This is seen in those attributes of the workforce beyond the classical functional and role fit, and this requires hiring managers to pay extra attention to how this learning value is embedded in the DNA of the staff that is joining the organization as a temporary or a permanent workforce.

Team dynamics remain a critical element to organizational health. There continues to be research and investment in peeling the onion on what the secret sauce is for the high-performing future workforce, yet the basics remain the same. Trust is a critical ingredient. Patrick Lencioni addressed the five dysfunctions of teams and how to overcome them. In his work, protecting the trust foundation requires the ability of the workforce to see healthy conflict as a positive quality that is necessary to build the future culture of strong ownership. Figure 1.4 reflects the highly digital team environment of the future. This is also an environment that will continue to struggle with finding the right balance between face-to-face and

Figure 1.4 Team Dynamics

virtual work environment thus leading most organizations of the future becoming increasingly accepting of hybrid being the modus operandi.

Driving toward excellence is another attribute of the future organization design. Powered by AI, organizations will continue to have the ability to make the strive toward excellence a dominating ingredient for how work is done and where time and energy are being spent. With the free time created by the proper use of technology, the future workforce has the ability to think again for a change.

This has a direct impact on the quality of work delivered and will ultimately contribute to the learning gained, which enhances the future project resource estimating. Excellence is a science and an art mix for marching toward the maturing of organizations and their project management practices. Being intentional in the future design in making this a priority and supporting it with the right culture, leadership, and workforce increases the likelihood of its achievement.

2

The Complexities of Project Workforce Estimating

LEARNING OBJECTIVES

- Understand the need for workforce estimating
- Plan for stakeholder involvement in staffing
- Understand workforce pricing structures
- Understand workforce estimating uncertainties and assumptions

Keywords *Direct costs; Indirect costs; Overhead costs; Portfolio management; Validating assumptions; Workforce cost estimation; Workforce databases*

Sources for Workforce Estimation

The requirements for a project can be something that may be so unique to the company that no estimating history exists. These types of projects are the most difficult to estimate. Fortunately, we always have some estimating data to work with, but we may find it necessary to seek out other sources of information. Typical sources of information include:

- **Lessons learned files:** Most line managers maintain lessons learned files related to estimating. When a project is completed, the functional managers review the hours and dollars for each element of work to see if their estimates need to be updated.
- **Estimating databases (internal):** Some companies maintain internal estimating databases as part of the firm's knowledge repository. This is quite common especially if the company has an estimating group responsible for pricing out most of the projects.

Project Workforce Estimating: Best Practices for Project Managers, First Edition.
Harold Kerzner and Al Zeitoun.

Companion website: www.wiley.com/go/Kerzner_ProjWFE

- **Estimating databases (external):** In some industries, such as construction, companies can purchase databases for estimating or subscribe to estimating services. These databases are often composite results from several companies and may be more reliable than internal databases.
- **Previous projects:** Analysis of previous projects often provides a good understanding of what the same activity may cost on the next project. This is particularly true if new equipment or work methods were instituted on the previous project thus making historical estimates no longer valid.
- **Subject matter experts:** These are people who have performed the tasks so often that they can estimate the activities with reasonable accuracy. The danger is that there may be too much reliance on these people and, if they were to leave the company, there could be a degradation in estimating.
- **Learning or improvement curves:** These curves state that the more often a worker performs a repetitive task, there is knowledge gained and the time to produce additional deliverables is usually less. This technique is appropriate for manufacturing organizations.

In addition to the above sources of workforce estimation, the future is bright with the use of artificial intelligence (AI). AI could address multiple areas of concern in the use of the above techniques. As an example, the bias that could be embedded in the estimate, due to various organizational and personal agendas, could be controlled. In addition, with the effective data analysis potential, there is room to look across larger amounts of data and capture trends and cross dependencies that might be missed by the human expertise.

All this adds up to a higher level of predictability and ensures the speed of reaching high-quality estimates. Agile practices over the years have also introduced additional simplified ways of estimating that use similar sources and that fit the nature of work for these types of projects. Ultimately, the goal is to find the most fitting sources for estimate that fit the organization, the complexity nature of projects, and the degree of risk that the project team is willing to take on.

Factors Influencing Workforce Estimation

There are two primary factors that influence project workforce estimating:

- Availability of the right resources
- Productivity level of the assigned workers

Worker availability is determined by the line managers. In non-project-driven organizations, project assignments may be of secondary importance to the functional manager than on-going work to support the daily business activities. In project-driven organizations, the functional manager may assign resources based upon the prioritization of the projects. In both types of organizations, the project manager may have the right to use contracted labor if qualified internal resources are not available.

Estimating productivity levels is quite difficult. Although there may be standards in place, there are things that can affect productivity. These include:

- Skill level or pay grade of the worker
- Availability of raw materials and the quality of the raw materials
- Availability of information (i.e. capability of the organizational process assets to provide the needed information)
- Quantity and quality of the desired deliverables
- Risk factors
- Safety measures
- Environmental factors such as the impact of unfavorable weather on construction projects
- Availability of equipment such as tools, jigs, cranes, and even permits

There are also personnel factors that can influence productivity such as:

- Breaks to use sanitary facilities
- Medical appointments
- Fatigue if extensive overtime is required
- Multi-tasking
- Motivation
- Breaks for coffee or smoking

In the workplaces of the future, and with the increase in the number of workforce generations working at the same time, this will add another degree of complexity to estimating the required workforce. Assumptions made about productivity or natural flow of work could be affected by this complex mix and cultural and background issues, and thus influence the ability of the project manager to handle staffing topics in such an environment.

TIP Factors influencing project workforce estimation will remain fluid into the future. As project work continues to change, so do the selection criteria, type, and number of staff.

The changing nature of work in the future and with organizations settling on a more hybrid model of work, could generate another set of factors that have to be addressed, such as:

- Ability to delegate
- Organizational politics
- Technology distractions
- Level of control
- Trust
- Style of leadership

In the future organization, a distinct shift toward outcomes and value achievement will affect the workforce design and contribute to refining the estimates as many of these factors are taken into consideration. This is where technology could help us create models that take much of this in the mix in order to potentially simplify how to reach a higher level of estimation accuracy.

Stakeholder Involvement in Workforce Staffing

In the early years of project management, students were taught to limit stakeholder involvement in projects for fear that they would meddle and create more problems than they could solve. This belief was based upon the notion that stakeholders possessed a limited knowledge of project management.

Today, stakeholder involvement is welcomed because of the knowledge they possess. Some stakeholders desire to participate in the staffing of the projects to ensure that the team members possess the desired skills. Stakeholders may have interfaced with some of the workforce on previous projects.

If stakeholder involvement in staffing is permitted, it should happen before the final cost of the project is determined. Stakeholders may request higher salaried team members to be assigned based upon their expertise. It is best to understand any additional staffing costs requested by stakeholders before the final contract is signed.

If managed well, the advantage of proper stakeholders' involvement in staffing is their strong buy-in. This could strengthen the selection of the right most fitting team and increase the chances for project success. Timeliness of that engagement and proper ongoing communication of staffing changes and the rationale behind them over the project lifecycle becomes necessary. In a world of increased project transparency, these kinds of shifts in the ways of working and collaborating could positively contribute to better project estimating if strategically managed by the project managers and sponsors.

The Key to Workforce Estimating

The work breakdown structure (WBS) and the earned value measurement system (EVMS) may very well be the two most important tools for the project manager regarding planning, estimating, scheduling, and controlling project workforce. When the EVMS was established in 1967, it was considered as the primary tool for integrating together cost, schedule, risk, and technical performance. It is therefore a management technique that relates resource planning to schedules, costs, and technical performance requirements. The WBS integrates everything together. Therefore, it is essential that the workforce be assigned according to the WBS elements of work.

EVMS emphasizes prevention over cure by identifying and resolving problems early. This includes problems related to poor project estimates. EVMS is an early warning system allowing for early identification of trends and variances from the plan. The EVMS provides an early warning of problems thus allowing the project manager sufficient time to make course corrections in SMALL INCREMENTS! The output of the EVMS includes:

- Measurement of resources consumed
- Measurement of status and accomplishments
- Comparison of measurements to projections and standards, including resource-hours
- Information serving as the basis for diagnosis, replanning, and possible rescheduling of resources

Once we obtain the output from the EVMS, we can answer the following questions and then re-evaluate whether we have the resources deployed properly:

- What is the status of the project?
- What are the problems?
- What can we do to fix the problems?
- What is the impact of each problem?
- What are the present and future risks?
- Can the assigned resources resolve these issues?

The EVMS tracks primarily time and cost. The tracking of cost can be done in hours, dollars, or both. Workforce tracking may have to be done separately although most software packages today do provide some information on resource deployment. Technology has helped in enhancing the powerful use of EVMS.

TIP WBS and EVMS continue to evolve in their value. Project workforce estimating could be both better planned and refined with the gained discipline and learning.

Earned Value Management Systems and the PMBOK® Guide

The ***PMBOK® Guide*** is an excellent source of information for project managers. The ***PMBOK® Guide*** provides information on planning, scheduling, and controlling project manpower.

Shown in Exhibit 2.1 are some of the activities related to manpower as they would appear in various ***PMBOK® Guide*** domain areas.

- **Initiation phase:** the executives select the person to become the project manager, and then the project manager negotiates with the functional managers for lead personnel. Executives and the project sponsor may exert their influence in having certain resources assigned to the projects.
- **Planning phase:** the project manager and the lead project personnel prepare the master plan and determine what additional functional resources, including possibly skill levels, are needed. Before the detailed plans are finalized, the project manager negotiates with the functional managers for the remaining manpower that will be needed.
- **Execution phase:** the project manager prepares the work authorization forms so that charge numbers can be established for the workers.
- **Monitoring and controlling phase:** the performance data is tracked for both dollars and hours. Discrepancies are analyzed, explained, corrected, and reported.
- **Closure phase:** the resources are released back to their functional areas for assignments on other projects.

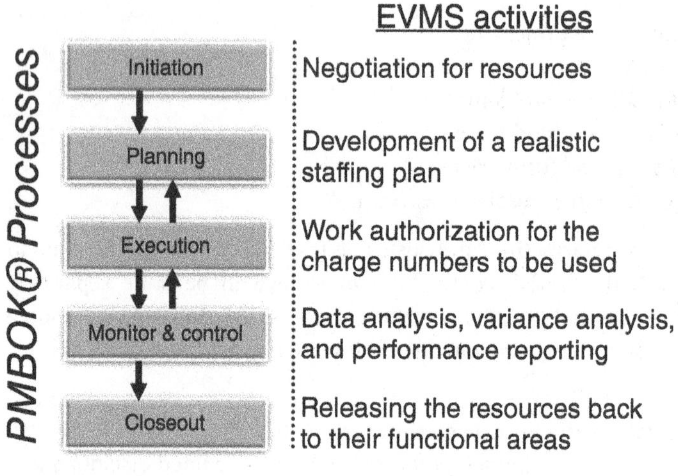

EVMS activities
PMBOK® Processes
Initiation
Negotiation for resources
Planning
Development of a realistic staffing plan
Execution
Work authorization for the charge numbers to be used
Monitor & control
Data analysis, variance analysis, and performance reporting
Closeout
Releasing the resources back to their functional areas

Exhibit 2.1 EVMS Activities

Direct Versus Indirect Project Costs

We have all heard the terms "fully loaded" or "fully burdened" costs. This means that, for every hour worked, there are indirect costs that must be applied as well as direct costs. Indirect costs are those costs that cannot be associated specifically with a common cost objective (i.e. project or program level) and must therefore be applied uniformly and consistently over that effort.

Consider the following example. You have been asked to perform an eight-hour task on a project. Your salary is $40/hour. Therefore, the project should be burdened $320 for the eight hours you worked. This is a direct labor charge against the project. But for the eight hours you worked, you were covered by the company's health plan, the company may have contributed to your retirement plan, and you were provided with a desktop computer, printer, phone, desk, and chair. These items are considered as indirect costs and are almost impossible to allocate to individual projects and programs, especially if the worker is sharing his/her time among several projects.

These items are typically identified as elements of the overhead of the company and measured as a percentage of direct labor. For example, if we say that the overhead is 150%, as identified in Exhibit 2.2, then for each hour worked, $40 is the direct labor cost and $60 is the indirect cost. The total or full loaded cost per hour is $100/hour, and the project will be charged $800 for the eight hours that you worked.

Another indirect cost could be the handling of raw materials you purchase on the project. The handling of raw materials ordered, tracking the materials, and maintaining the materials in inventory storage requires man-hours, but these man-hours may be priced out indirectly as a percentage of the cost of the procurement efforts. As an example, if you estimate that the cost of the raw materials will be $100,000, then you should budget for $100,000 plus another $4600 for material handling.

It is true that on some very large projects that have mega procurement activities, there may be full-time procurement personnel assigned to the project office. The critical decision is therefore the size of the project and the magnitude of the procurement activities.

Type	Method of Application	Example (%)
Overhead	A percentage of direct labor	150
General and Administrative (G&A)	A percentage of total cost	10
Material handling	A percentage of total goods	4.6

Exhibit 2.2 Indirect Costs

Breaking Down the Overhead Costs

Shown below are typical elements that make up the overhead of a company:

Building maintenance
Building rent
Cafeteria
Clerical
Clubs/associations
Consultants
auditing
Corporate salaries
Depreciation
Executive salaries
Fringe benefits
Group insurance
Holidays
Moving/storage expenses
Office supplies
Professional meetings Corporate
Retirement plans
Sick leave
Utilities
Vacations

Most companies do not have a standard overhead percentage applied to all man-hours. For example, manufacturing companies may have a very high overhead rate in the manufacturing division, perhaps as high as 500% or more because of all of the capital equipment that must be depreciated, whereas the engineering division may be burdened at 100%.

It is important to clearly understand overhead rates when staffing a project. Let's assume that your company has three pay grades for engineers and the hourly salaries are shown below:

	Salary	Overhead (%)	Burdened Salary
Junior engineer	$50	150	$125
Engineer	$60	150	$150
Senior engineer	$70	150	$175

Although this is a crude example, it shows that the senior engineer will cost your project $50/hour more than the junior engineer. Therefore, we can conclude that:

- Your project may not be able to afford to have the most experienced people assigned.
- If you wish to have experienced people assigned to a certain part of your project, then you may need to staff other parts of the project with average or below-average workers to maintain an overall reasonable average cost.
- During competitive bidding, estimating a project assuming that the best workers will be assigned may make you noncompetitive.

- Some projects are priced out using the average salary of the workers in a specific department. If, after the project go-ahead, you discover that the best workers were assigned, you must assume that the best workers can do the assignment in less time than the average workers and therefore reduce the number of hours allocated to that task such that the total cost will be the same. In the next section, we will discuss how adjustments may be necessary.

Let's go through an example on how the man-hours may need to be adjusted. We will assume that a junior engineer earns $50/hour and the hourly rate for a senior engineer is $80. With an overhead rate of 150%, the junior engineer is fully burdened at $125/hour and the senior engineer at $200 hourly.

Let's assume that, as part of competitive bidding, a work package was estimated at 2000 hours. The work was estimated for one junior engineer assigned full-time for one year. Therefore, this work package will cost 2000 hours × $125/hour, or $250,000.

If a senior engineer is assigned, then we must divide $250,000 by $200, which gives us 1250 hours. Therefore, the 2000 hours that were originally planned for may be reduced by 700–1300 hours. Of course, we are assuming that the senior engineer can perform the work in 1300 hours. If the senior engineer did the work in 2000 hours, then the cost overrun would be $150,000.

From this example, we can see how important it is to know the skill level or pay grade of the worker who will be assigned to perform the activity. Even with adjustments, there are always risks. A senior engineer may not be able to perform the work in significantly less time than a junior engineer. And using the above example, if the work package were estimated for a senior engineer, then assigning a lower-ranking engineer to perform the same work may require significantly more hours such that a cost overrun may be expected. In both cases, we see the importance of pricing work at the correct skill level. We can also assume that the use of software will help us in expediting the analysis of possible staffing scenarios to arrive at the right mix of grades, skills, and costs.

Forward Pricing Rates: Salary

Consider a company that has embarked upon a 12-month project with 6 months of work in 2022 and the remaining 6 months of work in 2023. When the work was first estimated, we knew the salary structure of the company for 2022. But in the first week of January each year, this company has a policy of giving out cost-of-living adjustments, promotions, bonuses, and other forms of salary increases.

Therefore, when pricing out a project that crosses multiple years, we must use forward pricing rates. Forward pricing rates identify what the company's best guess is for the salary structure for the next several years. Most forward pricing tables are for direct labor salaries and go out for three years. There are separate tables for projecting the overhead rates.

When pricing out manpower, it is now important to know in which year the resource-hours will be worked. When using forward pricing tables for workforce estimating, companies usually include a clause in the contract allowing for renegotiation of the remaining resource-hours if there is a significant difference between forecasted and actual salaries.

Forward pricing rates are obviously heavily oriented toward economic conditions in the host country where the work is taking place. Countries with relatively high inflation rates play havoc with forward pricing rates. Economic conditions can change rapidly. Forward pricing rates are used for salaries, overhead rates, and procurement costs.

Procurement estimating can be tricky. If you are working on a three-year contract and you know that the labor unions in the companies that supply you with raw materials will be renegotiating their contracts in the third year of your contract, the cost of your procurements can change significantly. All of these conditions are risks that have to be planned for and considered in project estimating and have a direct correlation with enhancing the estimates and the future ability of staying within the planned project budget.

Calculating Available Work Hours

It is important to understand how many hours an employee typically works each month without considering overtime or vacation. This often appears in the literature as the mythical man-month, and if not accounted for properly can result in significant cost overruns in labor.

As seen in Exhibit 2.3, a typical employee may not end up working 2080 hours/year on a project. If we take 2080 hours/year and divide it by 12 months, we end up with 173 hours/month. This is the mythical man-month because the employees do not work 173 hours/month. We must subtract vacation days, sick leave days,

Hours available per year (52 x 40):	2080 hours
Vacation (3 weeks):	–120 hours
Sick leave (3 days):	–24 hours
Paid holidays (11 days):	–88 hours
Jury duty (1 day):	–8 hours
	1840 hours

(1840 hours/year) ÷ (12 months) = 153 hours/month

Exhibit 2.3 Hours Available for Work

paid holidays, and possibly jury duty, when relevant. Now, the average employee works only 153 hours/month.

Sometimes this calculation is made per pay grade because some of the senior workers may have more vacation days available than the junior workers. Overtime can be included to adjust the resource-hours.

Work Authorization Form

As stated previously, resource-hours may have to be adjusted. Companies use work authorization forms that can be used to allocate and adjust workforce assigned to each work package in a work breakdown structure.

The form includes the hours to be worked in each cost center and the fully loaded cost of the hours. The form can also show that the resource-hours must be used between a given time period such as August 1–December 31. The people assigned to this work package will use a work order number that appears on the work authorization form when recording the hours worked. This work order number is open only during the time shown on the form. This pressures line managers to commit and assign the resources according to the plan.

The original work authorization form for a work package can undergo revisions, and the revision number also appears on the form. Larger versions of this form may also show how the resource-hours and costs change during each revision. The form may contain a description section that contains a complete description of all the work necessary to complete this work package. The description section may be identical to the description section in the WBS dictionary, where details of each work package are provided.

As ways of working continue to change, the increasing speed of authorizing work could create risks on the likelihood of remaining within the planned project estimate. The project manager will have to play a strong communicator role to get other stakeholders on board, update the initial authorization form, or escalate the need for project changes as necessary.

Project Pricing Overview

Pricing summaries for a project that would be part of competitive bidding are often more complicated based upon the information requested by the client. There are several key points:

- Departments or divisions can have different overhead rates. As an example, the engineering overhead rate might be 110% and the manufacturing overhead rate could be 200%.

- Corporate General and Administrative (G&A) is applied after all the labor and material costs are summarized.
- Profit is added on after corporate G&A is calculated.

Companies that do work with the Federal Government may create separate divisions dedicated to Government projects. For example, assume that the manufacturing division of a company produces products for both the Government and private industry. The products provided for private industry require the purchase of a $5 million piece of equipment.

It seems easy enough to include the depreciation on this piece of equipment in the overhead rate of the manufacturing department. But the Federal Government may complain that they are paying for equipment that has nothing to do with the products provided to the Government. The Government may then audit the company's overhead rates and ask them for a lower overhead rate for Government components. To simplify matters, companies usually create a separate Government Division if they expect to perform a great deal of work for the Government. This also includes the assigning of dedicated resources to work on Government projects to simplify workforce estimating.

Validating Estimation Assumptions

Projects very rarely have a go-ahead date that is immediately following the authorization of the project. Sometimes, the go-ahead date can be as much as six months or longer after the authorization and/or signing of the contract.

When projects are finally approved, there are usually assumptions made that are eventually documented in the project charter. The assumptions related to manpower include:

- People who will be assigned to the project will be at the pay grade that was used for estimating the labor costs.
- The forward pricing rates for labor are reasonable and probably accurate.
- The overhead rates for indirect costs will remain fixed as priced out in the contract.
- The forward pricing rates for indirect costs are reasonable and probably accurate.
- The labor we need will be available as planned.
- The amount of overtime included in the workforce pricing will either remain the same or be reduced if possible.

Once the project officially begins, the project manager must revalidate all these assumptions. The assumption that most often changes is the availability of labor.

The longer the delay between authorization and go-ahead, the greater the likelihood that the planned resources will either be assigned to other activities and are not available when needed thus resulting in a schedule delay, or other less qualified resources will be substituted for the planned resources. This can also be an area that results in customer dissatisfaction with the project team and the project progress.

Exhibit 2.4 illustrates a typical metric that can be used for tracking changes in assumptions.

As seen in Exhibit 2.4, in February there were nine assumptions made originally. Eight of the assumptions have remained the same, one assumption has been revised, and one new assumption has been added.

When assumptions on a project change, it is highly likely that the constraints on the project will change as well. Therefore, reiterations on the workforce planning model should also include a metric that tracks changes in constraints, especially critical constraints that may require a change in worker assignments. This is shown in Exhibit 2.5.

Successful execution of projects directly builds on the proper management of assumptions and constraints over the project lifecycle. The advantages of using AI in capturing and analyzing the fluid changes in assumptions and constraints are immense. The analysis of these data sets over time and across projects could result in critical learnings that affect future workforce estimating and that continues to increase work efficiencies.

TIP The disciplined management of assumptions and constraints contributes to increasing the likelihood of better estimates and contributes to maturing the organizational learning.

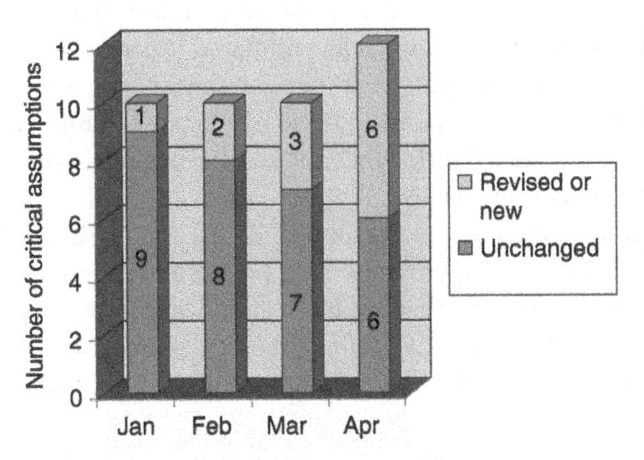

Month	Assumptions	
	New	Revised
Jan	0	1
Feb	1	1
Mar	2	1
Apr	5	1

Exhibit 2.4 Tracking of Assumptions

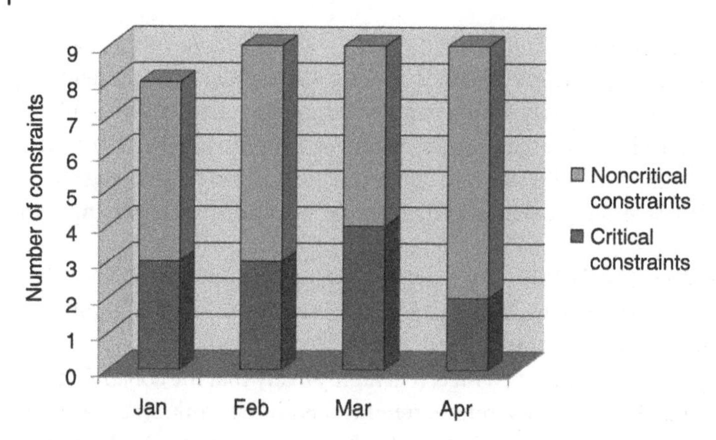

Exhibit 2.5 Critical Versus Non-Critical Constraints

The Fuzzy Front End

One of the biggest challenges occurs as to how projects are selected and prioritized. Historically, traditional project management practices began with often well-defined requirements provided in the business case or customer's statement of work. The skill level needed of the workforce team is then known, and the team usually has a clear picture of the deliverables right at the outset. Project management policies and procedures helped eliminate any uncertainties that may have existed by providing a structured process for products and deliverables, and the project's outcome was generally predictable even though an occasional surprise might appear.

Companies have recognized the benefits of successful implementation of project and program management practices and are now applying project management to projects that may not be as clearly defined as traditional projects at the onset. The new types of projects involve strategic initiatives, innovation, R&D, and creative thinking opportunities. Accompanying these new types of projects are new metrics, mainly business-related metrics, that relate to the business value created and how the firm can benefit financially.

On these types of projects, there often exists a great deal of uncertainty in the information needed initially to select and approve the right project and assign the correct priority. The uncertainty in the information has a significant impact on workforce planning and the end results. This "getting started" period, which occurs prior to actual product or deliverable development, can be highly chaotic and is called the "fuzzy front end" (FFE).

As highlighted in Figure 2.1, not all the required clarity will be present upfront. The amount of fuzziness increases depending on the nature of the project and

Figure 2.1 Clarity of the Front End

how much innovation is in the mix. The greater the fuzziness, the greater the uncertainty in the outcomes of the project and the expected business benefits and value. Unlike traditional projects that are highly structured and predictable, these types of projects are frequently unpredictable and unstructured, at least initially.

The FFE is not necessarily "Fuzzy." The FFE may be "Fuzzy" if it is based upon just an idea. The FFE may be somewhat more concrete if the focus is to determine the best way to capitalize on a breakthrough in technology that offers great promise. Regardless of the degree of fuzziness, the intent is to find ways to take advantage of opportunities.

There are several interacting participants in the FFE, each with varied skills. The participants must create, analyze, and evaluate many alternatives that could lead to strategic opportunities. Data-driven project management tools should be used in the FFE for better decision-making. Activities usually included in the FFE are:

- Market analyses
- Identification of internal strengths and weaknesses, and external opportunities and threats
- Other competitive factors
- Identification of potential customers
- Alignment to strategic business objectives
- Patents and copyright issues

The outcome of a successful FFE includes:

- Opportunity identification and analysis
- A product concept ready for implementation

- Identification of potential strategic partners and suppliers
- A business model aligned to strategic business objectives
- Preliminary product specifications
- Identification of sponsorship and governance personnel
- Criteria for product success and failure
- A startup action plan

With the number of modifications and changes that can occur during project execution on these types of possibly new strategic endeavors, executives must be sure that project problem-solving and decision-making by the workforce are in line with the intent of the FFE participants. Quite often, the information provided to the workforce to initiate the start of the project does not clearly identify the thought process or intent of the FFE participants. The result is that the project team may go off in the wrong direction.

TIP With the increasing uncertainty and volatility surrounding projects, managing the Fuzzy Front End is core to a risk management mindset that enables better outcomes.

Companies are now realizing that project managers, and potentially some of the workforce, should participate in the FFE activities to understand the concerns and intent. The knowledge that the project manager and workforce members gain by participating in FFE activities allows for better strategic and tactical decision-making during project execution.

Project Portfolio Management

When we discuss selecting a project during the FFE, the decision is often based upon what is in the best interest of the portfolio of projects rather than just focusing on an individual project. Project portfolio management (PPM) is the way that a company centralizes the control of all projects, even when the projects might not be related, under a portfolio of projects umbrella. This allows executives, managers, and stakeholders to make more informed decisions by seeing a big-picture view of all potential and ongoing projects that can impact the business. As companies take on more projects, the likelihood increases that decisions on one potentially new project could impact other projects in the portfolio. This occurs frequently with competing resources. Activities performed by PPM include:

- Strategic risk management
- Validating effective use of resources

- Determining the number of projects that can be undertaken
- Validating financial investment
- Portfolio problem solving
- Looking for ways to accelerate time-to-market for some projects

PPM requires the establishment of criteria to help decide if a new project should be included in the portfolio. Criteria might include alignment to strategic business objectives and corporate values, importance to stakeholders, financial factors, urgency, and risks. Scoring models can then be established to assign numerical values to each criterion for comparing new projects to those within the portfolio. The scoring models measure, rank, and prioritize projects and identify alignment to strategic objectives.

Once scoring models are in place, and each project's level of importance is identified, projects can be ranked. There are both qualitative and quantitative techniques to assist in the ranking process. This is important when considering the best allocation of resources. The ranking process need not be overly complex. It should allow you to monitor and adjust the portfolio as needed to cope with risks, uncertainties, and changes. The results of the scoring models are then used to determine the best allocation of employees to each project's workforce. Therefore, even though you consider your project as a high priority, it may be in the best interest of the portfolio for the resources you want to be assigned to other projects.

In addition to effective prioritization of projects and allocation of critical resources, PPM also makes it easy for executives and stakeholders to buy into the process and make informed decisions in a timely manner based upon evidence and facts rather than purely intuition. PPM requires project managers to use portfolio metrics to track business value, resource utilization, risks, and strategic metrics in addition to time, cost, and scope performance.

Project teams must understand that, as organizations grow, there can be multiple portfolios such as a strategic portfolio, international portfolio, IT portfolio, manufacturing portfolio, and various functional portfolios. Based upon priorities, environmental risks, and changing priorities, all projects are subject to unexpected changes in their workforce.

The Value Proposition Behind Project Portfolio Software Tools

Given the changing priorities, driven by new strategic choices that the executive team and potentially a governing board are making, it is important that all project managers have at their disposal the right PPM software tools that reflect the impact of these unexpected changes on their workforce planning. Project managers should be in a position to make concrete recommendations regarding the impact of these changes on the resourcing model, especially in iterative planning cases. It is becoming more common that project plans are not set in stone for the entire project journey and that a more hybrid mix of waterfall and agile planning tends to prevail. This makes it more difficult for the project manager in the future to control the accuracy of their workforce estimates.

Multiple PPM tools have been spreading in the marketplace. There remains a gap between the level of project and portfolio management practices and maturity levels and the vast functionality offerings that these tools provide. Many of these software tools have now included AI capabilities and sophisticated copilots that could literally make it extremely simple for the project manager to assess the health of the portfolio at any stage of the project from inception to completion. This could also highlight workforce gaps, future deviation predictions, and build the learning muscle necessary for strengthening the future workforce estimating capability.

The value proposition of using a PPM software, like Microsoft Project, could cover multiple key points, such as:

- Resource Allocation Optimization – leading to avoiding overallocation or underallocation and proper workforce fit
- Data-driven decision-making – critical for the future organization
- Enhanced Enterprise level Risk Management – key to the proactive way of working that supports future workforce planning
- Excellence in project governance – which supports the achievement of project value from the workforce

It is critical that investment in PPM software selection is based on a thorough understanding of the needs of governance for the organization and the project teams. This has to link to the level of infrastructure readiness that exists in support of this investment. Proper use cases, coupled with strong and well-planned training offerings, would increase the chances of the PPM software rollout success.

TIP PPM software tools are becoming a critical strategic differentiator for the project manager. Future workforce planning is educated by the rich data insights.

3

Techniques for Estimating Project Workforce Needs

LEARNING OBJECTIVES

- Understand workforce estimating techniques
- Identify the types of estimates
- Identify the hidden costs

Keywords *Backup planning; Dollar conversion rates; Enterprise risk management; Estimating methods; Hidden costs; Workforce needs*

Overview of Workforce Estimation Methods

There are several types of estimates. However, for simplicity's sake, we can consider that there are three generic categories of estimates:

- **Preliminary estimates:** Preliminary estimates are made from limited information, such as the general description of the project, the business case for the project or preliminary plans, and specifications that may have little or no detail. Preliminary estimates are used to get the project approved and into the portfolio of projects. Management must understand that, when approving a project using preliminary estimates, the final cost will most likely change. Preliminary estimates may not be made by the people who will eventually be required to perform the work.
- **Milestone planning estimates:** Once the project is kicked off, high level or summary level or milestone level planning takes place. These plans are prepared at the high levels of the work breakdown structure (WBS) and will involve some of the lead people from the various functional areas.

Project Workforce Estimating: Best Practices for Project Managers, First Edition.
Harold Kerzner and Al Zeitoun.

Companion website: www.wiley.com/go/Kerzner_ProjWFE

- **Detailed or activity estimates:** If the milestone planning estimates are significantly different from the preliminary estimates, management approval may be required for the project to continue. Assuming that management allows the project to continue, the next step is to prepare detailed plans and detailed estimates. This is normally done at the work package level of the WBS. Detailed estimating begins with a listing of all the steps required for a given project. If the detailed estimates are significantly different from the milestone estimates, management approval may once again be needed for the project to continue.

For companies that survive on competitive bidding, they may not have the luxury of going back to their own management to get approval to continue with the project. Based upon the type of contract, the estimate provided in the proposal may be the agreed-upon final contract price. In this case, manpower adjustments must be made to fit the final price rather than adjusting the final price to fit the manpower.

From Labor Hours to Labor Costs

Labor is usually first estimated in man-hours. Most standards for work are identified in man-hours. Then the resource-hours must be converted to dollars. There are three common approaches for converting hours to dollars. In the first approach, people are assigned to the project and then the hours are converted to dollars based upon the actual fully burdened salary of the workers who will be performing the work. This is probably the most accurate way of converting hours to dollars but does require knowing in advance who will be assigned.

Unfortunately, functional managers are not always capable of committing to who will be assigned to the project especially if there is a long delay between project approval and go-ahead. A similar problem is when certain resources may be required months after the project starts. This leads us into the second approach, which is converting hours to dollars using a weighted, fully burdened departmental average salary. This works well if the salary difference between the top and bottom pay grades in the department is reasonably small.

If there is a significant difference between the top and bottom pay grades, then it may be advisable to use a weighted, fully burdened departmental pay grade. If there are six pay grades, then there will be six weighted average salaries. Using this technique works well if the department manager is willing to commit to a pay grade assignment rather than the assignment of a specific person.

Enhancing Estimation Accuracy

If you know the amount of effort required for a project, you can estimate the total cost for the project and the associated time. However, the conversion from effort to dollars and time comes with inherent risks as discussed previously. The degree of error inherent to estimating is sharply reduced when work packages are well-defined. A well-defined work package is one in which activities are broken down in sufficient depth such that the chances of misinterpretation of what is required is extremely small. The better defined (detailed) a work package is, the smaller the degree of error in estimating, therefore a higher quality estimate. An acceptable range of error for a detailed work package is about 10%. On the other hand, when work packages are "guesstimated" without much detail, the degree of error is very high. The error range could span from 25% up to 100% and more.

But even with accurate estimates, other factors such as overtime and fatigue can increase the risks in the estimate. Not all workers at the same pay grade have the same level of efficiency.

Traditionally, workforce estimation has been done using estimating manuals and departmental standards. These estimates provide an estimate of the mean of workforce requirements but provide no indication of the spread about the mean. It is important to know the envelope or spread about the mean to truly understand the estimate and its accompanying accuracy.

TIP With advancement in artificial intelligence and data analytics, most barriers to estimation accuracy will be minimized.

Estimating Costs Per Hour

In a perfect world, we would know in advance which experts would be assigned to our project and we would then use expert judgment as the primary estimating method. Workers tend to know how long it will take them to perform a task. Unfortunately, we do not live in a perfect world, and we do not know in advance who will be assigned to an activity after it is approved and scheduled to begin. We must use other methods.

Although there are several methods for manpower estimating, the three most common methods are shown in Exhibit 3.1. The method chosen is based upon the

Estimating method	Generic type	WBS relationship	Accuracy	Time to prepare
Parametric	ROM	Top down	–25% to +75%	Days
Analogy	Budget	Top down	–10% to +25%	Weeks
Engineering (grass roots)	Definitive	Bottom up	–5% to +10%	Months

ROM = Rough order of magnitude

Exhibit 3.1 Traditional Estimating Techniques

complexity of the task, the risks, the availability of the workers (if known in advance), and the time and money available for estimating. Each estimating method comes with its own accuracy. The accuracy column in Exhibit 3.1 may be reflective of just one industry. The accuracies shown for a parametric estimate may reflect the construction industry whereas in IT the accuracy may be 150% to +200%.

Parametric Estimating

Parametric estimating is based upon statistics. For example, a contractor estimates that, based upon typical labor rates in the community, the construction cost of a home is approximately $125/square foot. Therefore, if the builder knows that you want to build a 4000-square-foot home, the construction cost (excluding the cost of the land) would be approximately $500,000. If you live in a high cost of living area, the cost per square foot might be $200/square foot in which case the construction cost will be $800,000.

Although parametric estimates are based upon statistical data, they can be inaccurate and lead to large cost overruns. Parametric estimates are usually made at the top of the WBS and, as such, are often referred to as "quick and dirty" estimates.

Parametric estimates are often used in competitive bidding when the company knows that the chances of winning the contract may be low and the company does not want to incur large bidding costs. The company may respond to an initial request for quotation with a parametric estimate and then use a different estimating method if the client asks several of the low bidders for a formal proposal.

Parametric estimates in the construction industry are quite accurate because of the volume of information that has been accumulated over the years. In IT, where we seem to come up with new computer languages every 5–10 years or less, parametric estimates may be highly inaccurate.

TIP Parametric estimating provides a valuable start at an accurate estimate that could be refined well as more WBS details are uncovered.

Analogy Estimating

Analogy estimating, also called budgetary estimating, is like parametric estimating in that they both are made at the top levels of the WBS. However, with analogy estimating, the estimator can adjust the standard based upon the degree of difficulty factor. As an example, a company has a standard estimate of 600 hours to perform a task. The estimator, or subject matter expert, believes that this task is 25% more complex than the standard task and therefore uses 750 hours as the estimate.

Analogy estimating is the most common form of estimating but is highly dependent on the skill of the person doing the estimating to determine the degree of difficulty factor. Analogy estimating is more accurate than parametric estimating and is used during competitive bidding efforts. The inaccuracies in the analogy estimates are usually the same as the profit percentage added into the proposal. If the analogy estimate can be off by 15%, the company may try to negotiate a 15% (or higher) profit margin in the contract.

Analogy estimates require the development of standards as a starting point in the estimating process. Some standards are based upon the man-hours needed in a worst-case scenario whereas other standards might be based upon a best-case scenario.

TIP In the future of project workforce estimating, technology could enhance the likelihood of a more accurate budgeting starting point.

Ground-up (Grassroots) Estimation

The most accurate workforce estimating method is "grass roots," bottom-up, or engineering estimating. These estimates are made at the lower levels of the WBS and are highly accurate. The disadvantage of this technique is that months or even years may be required to estimate a project, thus making estimating a cost endeavor.

A small Midwest construction company decided to give all their clients the best possible estimates. They priced out all jobs using detailed drawings at level 5 of the WBS. Unfortunately, they won only one job out of seven and eventually went out of business because of their estimating process. They could never recover their bidding costs.

Some companies maintain a bid and proposal (B and P) budget to bid on large complex projects that may require this amount of detail. In the aerospace and defense industries, it is common for some large projects to require years to be bid on. There could be hundreds of people involved in the estimating of one large job because the estimating is usually done from detailed blueprints rather than just sketches.

One of the reasons for the high degree of accuracy of these estimates is that the projects are normally large enough such that resources can be assigned full-time for the duration of the project.

TIP In the future, risks stemming from the high level of efforts in engineering estimating could be mitigated with technologies that continue to bring the real and virtual together.

Applying Learning Curves in Workforce Estimation

In manufacturing organizations, or companies that have a great many tasks that are labor-intensive and repetitive, learning curves or experience curves can be used. The principles behind a learning curve state[1]:

- The time required to perform a task decreases as the task is repeated.
- The amount of improvement decreases as more units are produced.
- The rate of improvement has sufficient consistency to allow its use as a prediction tool and a basis for estimating.

TIP Use of learning curves for workforce estimating requires a mature leaning organization culture and a commitment to the role of data in planning.

The learning curve in Exhibit 3.2 is a hyperbolic function. It implies that learning can go on forever, but that is not the case. The curve is based upon:

- A statistically derived relationship between the pre-production unit hours and first-unit hours that can be applied to the actual hours from the pre-production phase.
- A cost estimating relationship (CER) for first-unit cost based on physical or performance parameters can be used to develop a first-unit cost estimate.
- The slope and the point at which the curve and the labor standard value converge are known. In this case, a unit-one value can be determined. This is accomplished by dividing the labor standard by the appropriate unit value.

Learning curves are hyperbolic functions. But when hyperbolic functions are drawn on log–log paper, they appear as straight lines. Previously we said that, with learning curves, the rate of improvement is relatively constant every time production doubles. Therefore, if you have an 80% learning curve, then if the 200th unit requires 1000 hours, then the 400th unit would require 80% of the time required for the 200th unit, or 800 hours.

---

1 The learning curve material has been adapted from Kerzner, H. (2022). Project Management: A Systems Approach to Planning, Scheduling and Controlling, 13. Hoboken, NJ: John Wiley and Sons; Chapter 18.

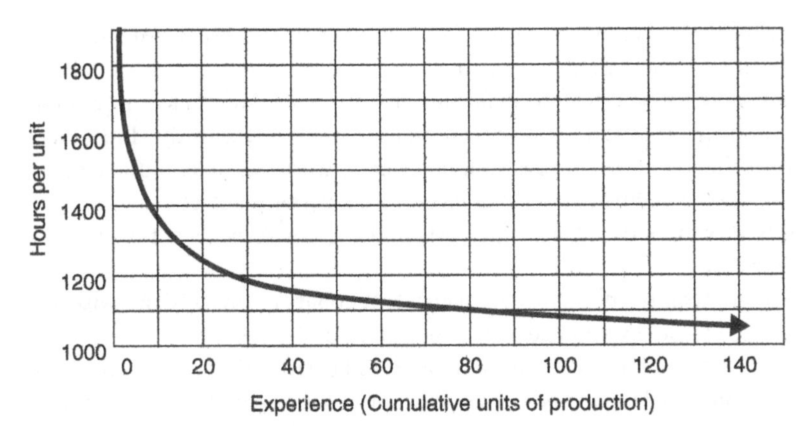

Exhibit 3.2 Typical Learning Curve

In most cases, the *Y*-axis is hours rather than dollars because dollars can change because of salary adjustments.

Learning curves are for labor-intensive efforts because labor can learn. There are several sources of learning to support the usefulness of these curves:

- Labor begins working with more efficiency
- Work specialization and methods improvements
- Introduction of new production processes
- Getting better performance from production equipment
- Changes in the resource mix
- Product standardization
- Product redesign
- Incentives and disincentives for performance

Understanding the Learning Curve Effect

In Exhibit 3.2, the learning curve was shown as a thin line. In reality, because of the differences in the way that people learn, the actual learning curve is a thick line, or what some people call a learning curve envelope. As an example, the hours needed for the 20th unit could be between 200 and 300 hours based upon the grade level of the worker to be assigned.

The envelope is a limitation to the use of learning curves. Other limitations include:

- The learning curve does not continue forever. The percentage decline in hours/dollars diminishes over time.

- The learning curve knowledge gained on one product may not be extendable to other products unless there exist shared experiences.
- Cost data may not be readily available to construct a meaningful learning curve.
- Other problems can occur if overhead costs are included with the direct labor cost, or if the accounting codes cannot separate work packages sufficiently to identify those elements that truly demonstrate experience effects.
- Quantity discounts can distort the costs and the perceived benefits of learning curves.
- Inflation effects must be expressed in constant dollars or hours. Otherwise, the gains realized from experience may be neutralized.
- Learning curves are most useful on long-term horizons (i.e. years). On short-term horizons, the benefits perceived may not be the result of learning curves.

TIP To enhance the effect of learning curve, technology, such as simulation, could take into effect the distinct differences across product conditions.

Estimating Management and Support Needs

Small projects are managed by a single person who is normally appointed by someone from the upper echelons of management. Since the salary of the project manager is known, estimating the management support costs is not complex even if the project manager is part-time on the project. Some companies have the project managers assigned as indirect costs paid out of overhead, but this practice is not recommended.

On larger projects, management support may be all the people assigned to a project management office. This includes the project manager and all of the deputy or assistant project managers. In this case, the project manager would be full-time but the assistant project managers could be part-time.

During the project selection and approval process it may be impossible to know the size and composition of the management support needs. As shown in Figure 3.1, and as a rule of thumb, management support for labor-intensive projects is generally 12–15% of the total labor needed for the project. If 20,000 hours of labor are needed on a project, then 3000 hours of management support may be necessary. For capital-intensive projects, such as installing a new piece of capital equipment, management support is about 8–10% of total labor.

The management support cost is heavily dependent on the complexity and risks of the project as well as the organization's project management maturity level. In mature organizations where workers (including line managers) are reasonably mature in project management, the management support costs may be less.

Figure 3.1 Management Support Rules of Thumb

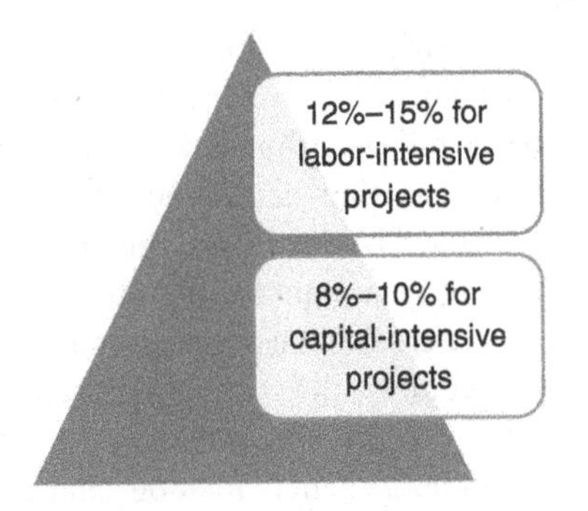

In some companies, management support costs also include the cost for line managers and executives to attend project review meetings on a periodic basis. Some companies allow functional managers to charge one or two hours a week as a direct labor charge against the project to attend project review meetings and other project-related work such as supervision of their workers who are assigned to the project. Executives may also charge two hours a week to the project if they are functioning as a project sponsor. This occurs most frequently on larger projects that are being performed for external clients.

Identifying Hidden Labor Costs

There are often hidden costs involving manpower that suddenly appear on a project. This happens because, as a project manager, we must expect the unexpected. When problems occur, the result is normally a meeting or a series of meetings. There are many reasons why escalations take place regardless of how well planning was conducted.

Brainstorming meetings usually result in hidden costs. On some projects, there can be a need for several brainstorming meetings. Quite often, these meetings involve workers who may not be part of the project team but may have valuable ideas and possible solutions to problems. These individuals may ask for charge numbers against which to allocate their time. As the ways of working continue to change to a more collaborative approach, there is likely a wider use of brainstorming, even virtually, and the implications of that shift have to be considered in workforce estimating.

Another example is when employees have personal issues, such as with an employee performing in a toxic manner. These employees may require counseling sessions with Human Resource Department personnel and use their charge number for billing their time in the meetings.

If the meetings are held away from your company, then there may be additional costs for airfare, meals, lodgings, and so forth. But the real hidden cost might be the salaries of the people who must accompany you to the meeting. In general, anybody accompanying you to the meeting will ask for a direct labor charge number to bill their time against your project. If two managers are accompanying you to the meeting, and for simplicity's sake we assume that the total time including travel will be three days, then the cost could be significant. If the salary of the two managers accompanying you is $200/hour fully burdened, then the additional charge against the project might be $9600 plus travel expenses. If several meetings are required, the costs could become quite large.

Some companies establish travel budgets as part of project cost estimating. The travel budgets may include salaries as well as travel costs. Even though the contract may be a firm-fixed-price effort, the travel budget may be treated separately as a cost-reimbursable add-on.

This topic of hidden workforce costs is becoming a critical one to include in the project manger's risk management practices. With the improvements in learning across projects, there is a better chance that the impact of such hidden costs will be properly addressed in project contingencies or that the project manager will have some pre-planned actions to be creative with how these costs could be handled.

The Impact of Documentation on Labor Costs

It would certainly be nice if we could achieve paperless project management. Unfortunately, even with dashboard reporting efforts, other reports and handouts for a meeting remain necessary.

The steps needed to prepare a report include:

- Organizing the report
- Writing
- Typing
- Proofing
- Editing
- Retyping
- Graphics Arts
- Approvals
- Reproduction
- Distribution
- Classification
- Storage
- Disposal

These steps take time. Some companies estimate that between eight and ten hours per page are needed to perform all these steps, including everyone who may be involved.

It is imperative that, when negotiating for resources for the project, you have a clear identification of what reports are needed so that you can negotiate for people who have writing skills. Getting to the end of a project and discovering that you have workers who lack writing skills can pose a problem.

It is also important to price out the man-hours needed to prepare the reports. This is another reason for clearly knowing the documentation requirements of a project. If a worker tells you that they need eight hours to perform a test, then you might allocate eight hours for the worker in the budget. Later, you discover that the worker charged 24 hours against your project because nobody included the 16 hours that were needed to write up the results of the tests.

This category of workforce efforts is one of the ripest for technology disruption. With the continual advances in artificial intelligence (AI), generative AI could not only take care of most of the report writing work but also help in filling the writing skills gaps that might exist in the project team. This massive change in how project teams work, will not only lead to much higher efficiencies but also result in an updated view of where the project manager and project team spend their time. The project workforce might become visible again.

The Need for Workforce Backup Plans

We all run the risk that something unforeseen may happen to one or more of our critical resources. Sometimes resources are removed from our project immediately to help put out fires elsewhere in the company. Sometimes people resign and leave the company immediately. Other times, people simply get sick or get hurt, and we end up with no qualified replacement.

Most well-managed companies develop succession plans for people in management slots. Each manager is expected to have someone in their organization ready to fill their position should they get transferred, promoted, or become ill. Years ago, large Government programs overfunded the program management offices that were providing governance for the programs. People assigned to the program management offices were expected to serve as a backup for one or more program office workers should anything bad happen. The Government recognized this as an over-management cost and was willing to incur the costs.

As reflected by Figure 3.2, backup planning requires analysis of multi-data pieces and developing and using a holistic approach to look at the project end-to-end in order to come up with good and diverse recommendations. Most project teams are running lean and mean. Functional organizations support training programs for resources and subject matter experts such that more than one person is qualified to fill a position. The learning curve for replacements in the functional ranks may be low. However, based upon the size of the project, it may be advisable to have one or more assistant project managers assigned who can fill the shoes of

Figure 3.2 Criticality of Backup Planning

the project manager in an emergency. The assistant project managers may be part-time rather than full-time workers on the project. Obviously, the size, risk, and complexity of the project are the determining factors.

Today, we are working on projects that are longer than before and more strategic in nature. Allowing these projects to fail could be quite costly. The solution is now in the development of backup plans for key project workforce personnel as well as individuals in key management positions. Backup planning is now becoming a critical component of workforce planning.

Common Challenges in Workforce Estimation

As you can surmise, there are numerous manpower estimating problems that can occur on a project. Some of these problems include:

- Poor manpower estimating techniques and/or standards, resulting in unrealistic budgets
- Out-of-sequence starting and completion of activities and events that may create problems when resources should be assigned to the project or leave the project
- Inadequate WBS that makes it difficult to accurately determine the hours needed for a work package
- Management reducing budgets or bids to be competitive or to eliminate "fat," and you are asked to slash your manpower estimates or use lower-salaried workers

- Inadequate formal planning that results in unnoticed, or often uncontrolled, increases in scope of effort, thus requiring additional man-hours
- Unforeseen technical problems resulting in the need for more man-hours
- Schedule delays that require unplanned overtime or idle time costing
- Failure to understand customer requirements
- Unrealistic appraisal of in-house capabilities of the workers
- Misinterpretation of information such that more work is actually needed
- Use of wrong estimating techniques for manpower estimating
- Failure to assess and provide for risks in the manpower estimates
- Accepting customer requests for changes or additional work without management approval
- Having scope changes approved but not having sufficient manpower to work on the scope changes

The Essential Value of Enterprise Risk Management

Future project workforce estimating excellence could build on a foundation of advanced risk management practices. Risk management is the cornerstone of proper project management. With the critical proactivity shift that project managers esteem to look for, risk management serves that need. By nature, risk is due to uncertainty, and it is about something that has not yet happened. In the case of workforce estimating, the project manager's ability to predict what to do to handle uncertainty and complexity in estimating future workforce needs, will depend on the consistency of utilizing enterprise risk manager. The enterprise element is a unique advantage.

The enterprise view of risk management enables a holistic understanding of the multiple factors that come together affecting the ability to properly estimate workforce needs. These factors include multi-tasking, multiple moving parts within the project portfolio, political agendas, leadership behavioral consequences, and are all examples for the need to perform enterprise risk management.

Enterprise risk management (ERM) is usually present in organizations that have matured in their practice of project and program management. Naturally, these organizations would form an ERM committee that would include a few of the key executives, such as the Chief Financial Officer (CFO) and several others. This committee's focus is ensuring that the right dialogs and proper level of stakeholder involvement in proactive project planning and control activities take place. This could directly contribute to better estimates upfront and an active engagement in the workforce estimating refinements that need to continue to take place.

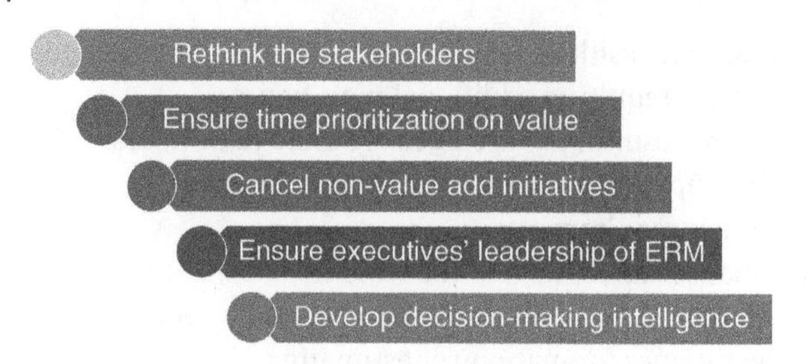

Figure 3.3 Enhancing Estimating with ERM

ERM looks at risk both functionally and organizationally. The organizational piece combines culture, leadership, and supporting enablers. Although the management involvement in this committee could be yet another item on the list of what should be estimated, as was highlighted in Figure 3.1, it is an investment that would pay off in reaching the achievement of the right valuable outcomes from planning for the right skills and numbers of workforce.

An important success element in the success of ERM is having the right people in the room and running the right open tough dialogs. The art of proper workforce estimating in the future hinges on our ability to see risk management and the potential projects and extended environment uncertainties as opportunities to better learn and use that learning to make better decisions around our estimates and the related risk handling strategies. Figure 3.3 is an opportunity to rethink the stakeholders who need to be involved in ERM and having more of a meta view of who to include in the ERM activities, which ultimately contributes to enhancements in planning and estimating. The value angle is also a critical future shift. When we estimate the workforce, we are going to focus on the value expected out of the properly developed WBS, or at least in the case of agile, a nicely prioritized backlog that is value centered.

Holistically ERM could contribute to canceling non-value-added programs and projects. This will free up resources and focus their use in the most critical projects. This portfolio capability ensures that we are investing in the right projects and assigning our workforce accordingly. Achieving this level of practice maturity is linked to how closely the executives lead and the example they demonstrate. This is the new wave of project management where project success is refined as projects are seen more as strategic vehicles in the future. The last element of the figure is activated with the intelligence that data and technology enable in the future and thus elevates the impact of ERM in bringing a higher planning and workforce estimating quality.

4

Monitoring Workforce Expenditures

LEARNING OBJECTIVES
• Understand workforce tracking systems • Identify workforce metrics • Identify workforce reporting techniques

Keywords *Business models; Cost tracking; Documentation; Trend analysis; Workforce metrics*

Initiating Workforce Expenditure Tracking

One of the questions facing most project managers is when to begin tracking project manpower. There are several compelling reasons for wanting to begin the tracking process at the beginning of the project:

- The people assigned to the project are at a higher or lower pay grade than anticipated. The earlier this is discovered, the quicker the adjustments can be made.
- We have a shortage of qualified resources.
- We are using overtime right at the start of the project.
- The resources that were scheduled to work on a full-time basis are working only part-time.
- Critical resources have left the company.

In the early life cycle phases of a project, there are numerous opportunities to correct problems or head off disasters, and the cost of correcting the problems is

Project Workforce Estimating: Best Practices for Project Managers, First Edition.
Harold Kerzner and Al Zeitoun.

Companion website: www.wiley.com/go/Kerzner_ProjWFE

usually low. In the later life cycle phases, the opportunities to correct problems are significantly less, and the cost of corrections is large.

TIP Initiating workforce expenditure tracking earlier provides a strategic opportunity for enhanced management of the project budget.

Converting Work Hours into Financial Metrics

Unless we know in advance specifically which people will be assigned to a project, we can only make an educated guess as to what the salaries will be, and this is accomplished using a blended rate. The blended rate allows us to develop a cost baseline for the project. But after the workers begin performing work, how do we convert the actual hours worked to dollars? There are three choices:

- Work is initially priced out at the department average, and all work performed is charged to the project at the department average salary, regardless of who accomplished the work.
- Work is initially priced out at the department average, but all work performed is billed back to the project at the actual salary of those employees who perform the work.
- The work is initially priced out at the salary of those employees who will perform the work, and the cost is billed back the same way.

Reporting worker costs on an average salary is not good because, if the worker was earning more money than the average salary, you lost money and did not realize it. If the worker was earning less than the average salary, you may have made extra profit and did not realize it.

Allowing all hours worked to be reported using the full-burdened salary of the worker is best. However, some companies do not allow this to happen because the project manager could end up with the knowledge of the exact salary of all the workers on the project team. This example illustrates how important it is to track both hours and dollars.

TIP Advancements in digitalization could be an opportunity to address the concern with access to actual salaries and simplifying the transition from average to actual data.

Balancing Hours and Dollars in Project Tracking

Most people that use the earned value management system (EVMS), seem to prefer to track only dollars. Unfortunately, dollars alone cannot give you an accurate picture of how well the resources are being deployed. When analyzing resources, it is necessary to track both hours and dollars.

As an example, let's consider a work package where schedule variance is positive, and the cost variance is negative. This means that we are ahead of schedule but over budget. There are several possible causes for this, including:

- Use of overtime
- Use of higher-than-expected salaries or higher-than-planned for pay grades
- Additional resources were added to the work package
- Other causes

As you can see, we must dig deeper to find out the root cause of the variances, whether favorable or unfavorable. Now, let's report both hours and dollars. In this example, we will assume the cost variance in dollars is unfavorable, but the cost variance in hours is favorable. Therefore, we are spending more money with fewer hours, which implies that we are using higher-salaried workers than we originally planned on using.

As reflected by Figure 4.1, the topic of workforce tracking could additionally be more complicated when other factors are added to the mix. For example, changes in the work breakdown structure, different staggering of work, globalization impact, or higher utilization of technology. It is critical that the project manager remains vigilant in capturing and presenting an accurate picture of workforce utilization that supports proper and timely decision-making resulting in any necessary shifts.

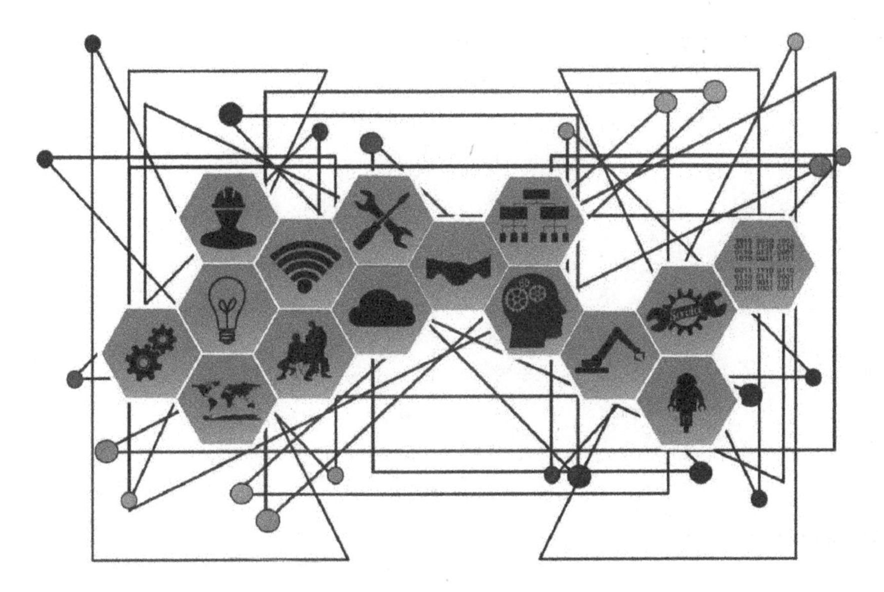

Figure 4.1 The Dynamic Nature of Workforce Tracking

Analyzing Workforce Metrics

In addition to cost metrics, it is also important to establish metrics that show the skill level of the workers assigned as well as the number of workers. Examples appear in Exhibits 4.1 and 4.2.

Exhibit 4.1 shows the number of resources actually assigned for each work package versus what was planned. For Work Package #1, five resources were planned, but only four are assigned. This could indicate a shortage of resources or that the four resources assigned are able to perform the work of the five resources that were planned.

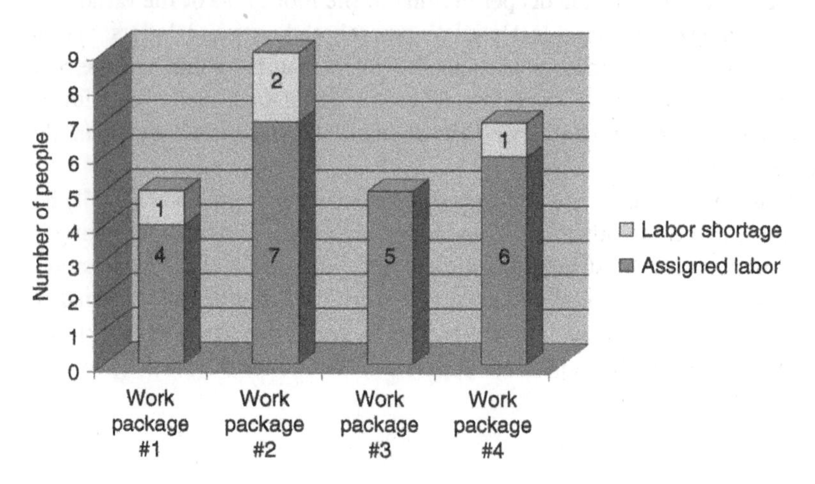

Exhibit 4.1 Assigned Versus Planned Resources

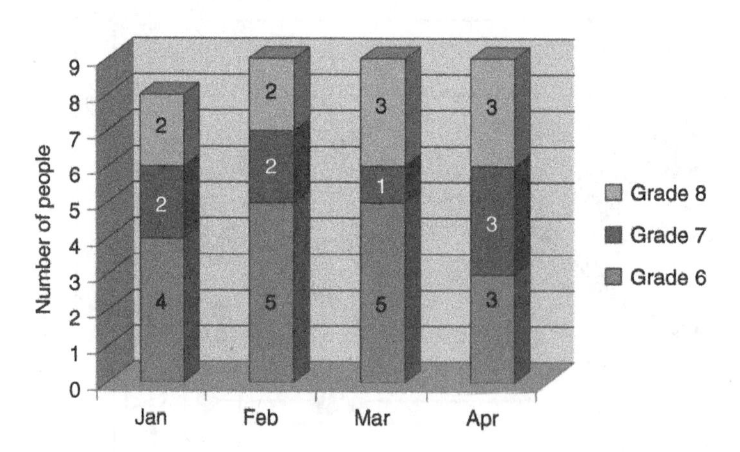

Exhibit 4.2 Grade Level of Assigned Resources

For effective workforce planning, it is important to understand the quality, or the grade level, of the resources assigned. This is shown in Exhibit 4.2.

As shown in Exhibit 4.2, nine people were assigned to the project in February. However, even though the workforce plan called for nine people to be assigned, the workers may not be at the correct pay grade. As an example, let's assume that a Grade 8 has better skills than a Grade 7, and a Grade 7 has better skills than a Grade 6. If the workforce plan for February required nine workers, then the headcount is correct. But if the workforce plan was based upon only Grade 7 and Grade 8 workers to be assigned, then five people are assigned with less than expected skills.

Some companies, such as construction organizations, employ salaried workers, hourly workers, and contracted workers. Each worker can have a different base salary or fully burdened salary. An example of the headcount metric might look like Exhibit 4.3.

Another important metric appears in Exhibit 4.4.

As shown in Exhibit 4.4, it is also important to know when the resources will be working on your project. Workers performing their activities overtime will most likely be paid more than when they work on regular time. Exhibit 4.4 also identifies whether there exist unstaffed hours.

Analyzing Spending Trends

Exhibit 4.5 illustrates a typical spending or *S*-curve for a project. The *Y*-axis could be dollars or hours. For status reporting to a client, it is usually dollars, but for manpower analysis, it could be hours or dollars.

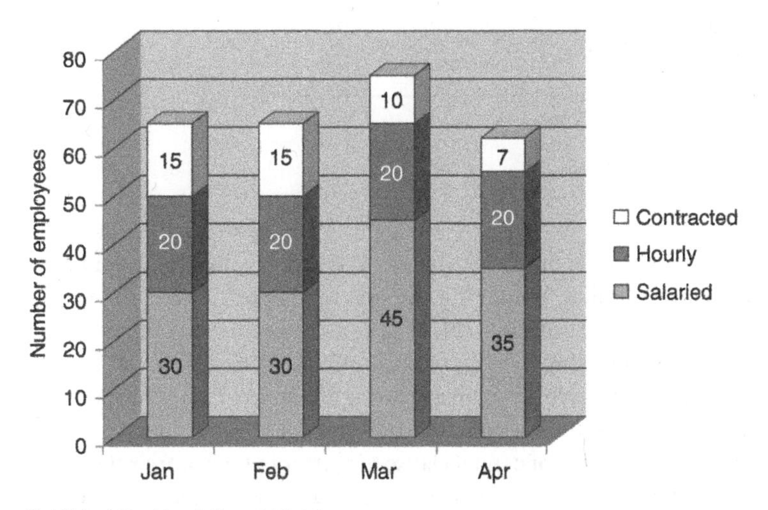

Exhibit 4.3 Head-Count Metric

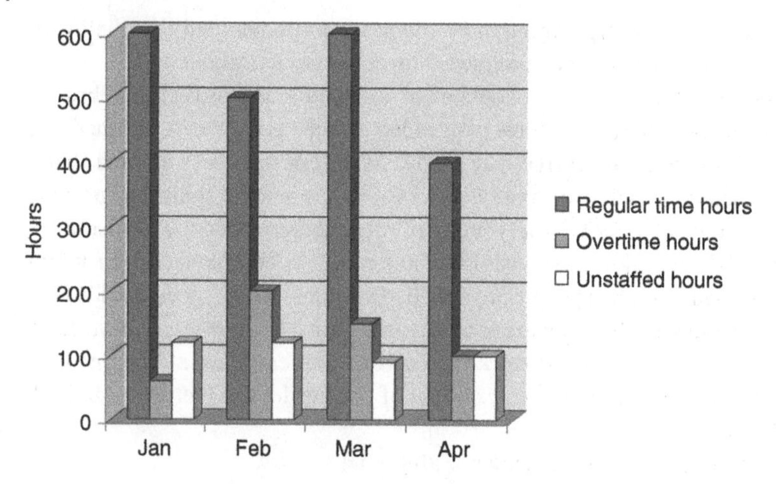

Exhibit 4.4 Regular, Overtime, and Unstaffed Hours

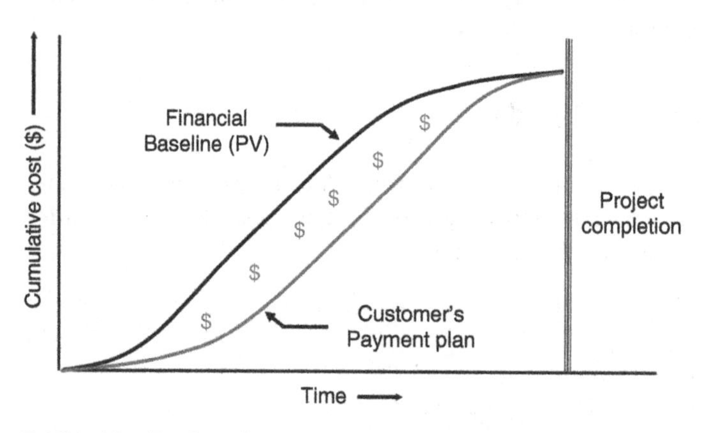

Exhibit 4.5 The Spending Curve

The financial baseline, or planned value of work, is the way the contractor plans on assigning resources and accumulating costs on the project. This is the way it will probably appear in the proposal if this is part of competitive bidding.

The customer's payment plan is the way that you will receive payment for the work performed. Since the payment plan appears after the work is performed, the contractor may have to use their own money to fund the project during the gap. Some contracts allow for a cost of capital clause to appear in the contract whereby the contractor is allowed to charge the client the interest on whatever money is borrowed to fill the gap.

If the contractor does not have the necessary funding to start the work according to the financial baseline, then the contractor may begin assigning the resources closer to the customer's payment plan than the financial baseline. There have been many lawsuits over this, especially when the end date of the project slips because of the delay in assigning resources when stated in the proposal.

TIP As part of proper risk management, the project leader should continuously analyze spending trends to mitigate the chances of legal trouble.

Example of Termination Liability

There are a multitude of reasons why projects get canceled prior to the planned completion date. When the project is internal to your company rather than for an external client, the resources simply return to their respective functional organizations, awaiting their next assignment. But if the project is being funded as contracted work for an external client, there may be a termination liability clause included in the contract.

Termination liability is the amount that the client owes the contractor for the privilege of canceling the contract prior to the regular completion date. Termination liability applies to manpower and procurement activities, but we will consider only labor.

Assume your company was awarded a six-month contract where the fully burdened labor costs were $50,000 each month for six months beginning the first of March. At the beginning of May, the client informs you that they wish to cancel the contract at the end of May. The termination liability fee in this example is 80% of the following month's labor. The rationalization for this is that you may have to lay people off, and this 80% would be part of their severance package or placing the workers temporarily in an overhead pool as they await their next assignment or seek employment elsewhere in the company.

In the above example, you will have spent $150,000 in labor through the end of May. Adding to this number 80% of June's labor cost, you can bill the client $190,000. Termination liability could have been 80% of the labor for the next two or three months rather than one month. This would happen if you hired contracted labor and had to agree to a minimum of several months of employment for the workers.

Oversight of Workforce Expenditures

Accuracy and honesty in resource-hour reporting is essential. Some contracts, such as cost-reimbursable contracts, allow the customer to audit the costs of the project periodically to make sure that all the charges are correct and billable. On large Government contracts, the Government may have auditors that may reside within the contractor's company for the duration of the contract. The auditors may even go around the company and audit personnel at random, asking them what charge numbers they are using and what projects they are working on.

There have been situations on Government contracts where the company would be running out of money on a firm-fixed-price contract and then set up fictitious charge numbers on another Government contract that was a cost-reimbursable type contract. The workers would use the fictitious charge numbers that would end up being billed against the cost-reimbursable contract when they were performing work on the firm-fixed-price contract that was almost out of funds. By the time that the fraud was detected, more than $5 billion in phony charges had accumulated.

As illustrated by Figure 4.2, it is critical that there is a proper level of oversight throughout the project lifecycle to handle outcomes of audits, internal changes, and possible customer's updated plans. Transparency in conducting the oversight is critical, and the use of advanced data analytics contributes to gaining more meaningful insights from the oversight efforts. We do have laws requiring truth of disclosure of information, especially financial information related to workforce charges.

Figure 4.2 Workforce Oversight

Setting Reporting Intervals for Workforce Status

Everybody recognizes the need for effective reporting of information, but most people simply do not know what information should be reported and how frequently. The information being reported depends on the type of project and characteristics of the organization. Project-driven organizations generally report status differently than non-project-driven organizations. Project-driven companies may report status daily or weekly, whereas non-project-driven companies report monthly.

The information provided in the report can vary based upon who will be receiving the report. In general, the information is provided as shown below. Even with sophisticated software, companies cannot calculate changes in the overhead rates on a daily basis. Some companies even do it just yearly. On a weekly basis, or even a daily basis, project managers are interested in the direct labor hours and dollars. While some people argue this, most project managers appear to control hours rather than dollars.

Weekly reporting:

- Direct labor hours
- Direct labor dollars

Monthly reporting:

- Direct labor hours
- Direct labor dollars
- Overhead and G&A
- Other expenditures
- Materials
- Variances (in hours, dollars, or percent)
- Cumulative-to-date data
- Payments made
- Profits booked
- Problems

TIP Sensitivity in selecting which workforce data gets reported and how frequently, is dependent on who receives the report and the degree of being project-based.

Documenting Challenges in Workforce Reporting

Most people that work on projects seem to treat reports as red tape that prevents them from doing their job. On some projects, people may spend as much as

25–30% of their time writing reports. Although we are trying to go to paperless project management, some reports are still needed.

The three reports commonly created are:

1) **Progress reports:** These reports indicate how many resource hours were spent on each one of the work packages during the reporting period.
2) **Status reports:** These reports tell us about the variances in resource hours between planned work and performed work. These are the man-hour variances and can be favorable or unfavorable.
3) **Forecast reports:** These reports indicate approximately how many resource-hours will be needed to complete the remaining work packages, whether they are still open or haven't started yet. This is important so that management knows when the resources will be freed up to work on other projects.

There is additional information that will appear in each report, but we are just focusing on resource hours.

Young and inexperienced project managers place too heavy a reliance on computer technology for status information. While computer technology can tell you the hours worked, it cannot tell you if the right resources were assigned or if the resources are doing the work correctly.

Computer printouts have value, but they are not a replacement for walk-the-halls management or other ways of directly pulsing what truly is happening with the project team. Talking to the people and seeing how they perform the work is the only real way of determining if the right resources were assigned. On large projects, this may be difficult to do.

TIP Having the most fitting workforce for the project's needs remains a cornerstone for what the project manager should focus on, rather than just purely the hours.

Optimizing the Business Models Around Workforce Strengths

Organizations exist for a reason. They provide products, solutions, and services that affect the sustainability agendas of the future. Whether the origination exists to achieve a profit or is a non-profit, the organization must develop its business model to align with its purpose and the capabilities of its workforce.

As seen in Figure 4.3, the dynamics of workforce strengths are changing fast. This is, in large part, due to the fast-growing enablement of the workforce with digitalization. The workforce in the future will be equipped in a way that makes it possible to reinvent the business models, given the increased capacity, the higher

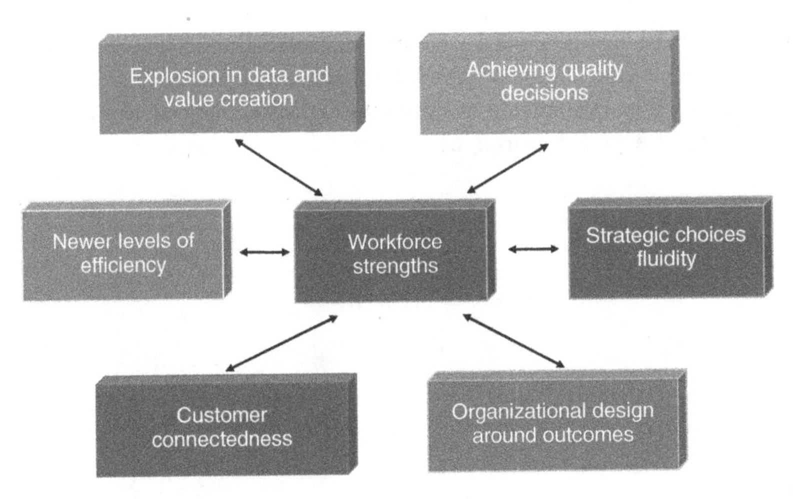

Figure 4.3 Business Model Considerations

possible efficiencies, shorter product life cycles, and the almost complete collapse between the real and the digital. This is a work revolution that has never been encountered, and it is taking place while we have about five generations of the workforce in the workplace collaborating at the same time.

The six areas resulting from enhanced workforce strengths shown in Figure 4.3 could collectively contribute to rethinking an organization's business model. Most of these areas are directly a reflection of enhancements resulting from ongoing digital transformation achievements. The world of work is transformed, and no matter the industry, the impact of technology on where the workforce spends its time, and why, is being questioned and is changing.

The massive amount of meaningful data and the relatively easy and economic access to it have opened the door for stronger collaboration, created a more open business ecosystem, allowed partners to work closely with organizations, and led to the introduction of many new startups. In the future, full-time and contract workforces will have more exciting work opportunities and will be able to use the improved decision-making capacity to have a more relevant impact on the work results. This more empowered future workforce will be able to reach strategic choices more fluidly and thus create flatter structures and business models that are enriched by being able to move fast to value.

Organizational design that is done in a more projectized way with a focus on outcomes will also support innovations in future business models. When this is coupled with higher connectedness with the customer, project workforce will develop new definitions for project success that serve more dynamic business models. These models will be anchored in cultures and a workforce that are able to adapt fast with the changing market conditions and the increasing customers' expectations. Much of the ability to rethink future business models is serviced by the new levels of efficiency that are achieved through technological innovations when well used by the project workforce.

5

Growth of Innovation Project Teams

LEARNING OBJECTIVES
• Identify innovation needs • Identify types of innovations • Understand principles of co-creation • Understand definitions of success and failure • Understand innovation value • Identify characteristics of an innovation culture • Understand the need for innovation portfolio analysis

Keywords *Agile; Innovation project management office; Need for innovation; Prompt engineering; Types of innovation*

The Need for Innovation and Creativity

Over the past three decades, there has been a great deal of literature published on innovation and innovation management. Companies need some degree of innovation for growth. Converting a creative idea into reality requires projects and some form of project management. Staffing must include workers that can think creatively. Unfortunately, innovation projects may not be able to be managed using the traditional project management philosophy we teach in our project management courses. We must be sure that the workers assigned have innovation capabilities. Innovation varies from industry to industry, and even companies within the same industry cannot come to an agreement on how innovation management should work. This adds complexity to staffing.

Project Workforce Estimating: Best Practices for Project Managers, First Edition.
Harold Kerzner and Al Zeitoun.

Companion website: www.wiley.com/go/Kerzner_ProjWFE

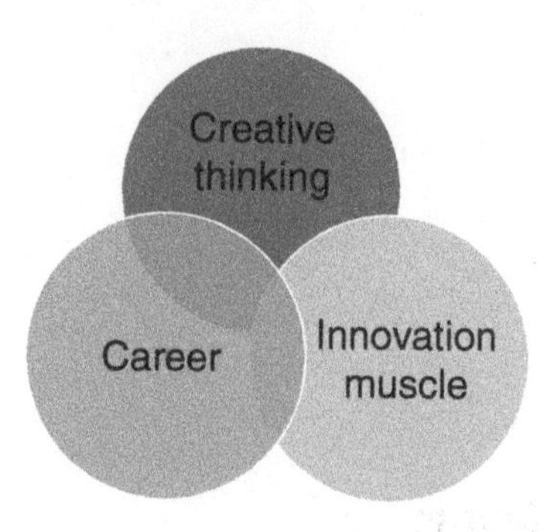

Figure 5.1 Innovation for Growth

It is inevitable that, over the next several years, professional organizations such as the Project Management Institute [PMI®, 2024] will recognize the need to begin setting some standards for staffing projects requiring innovation. This may appear as a certification program in innovation project management (IPM) or a series of courses. It may also appear as an IPM Manifesto like the Agile Manifesto.

This combination of ingredients to enhance the opportunity for innovation growth is summarized in Figure 5.1 and will require continued nurturing by the project manager and by the future's workforce. The greatest innovation in the next decade may be the recognition and advancement of IPM as a career. The intent of this chapter is to identify several of the differences between traditional and IPM along with the accompanying staffing challenges and to provide the basis for understanding the need for some degree of standardization in IPM.

Introduction to Innovation

Companies need growth for survival. Companies cannot grow simply through cost reduction and reengineering efforts. Also, companies are recognizing that brand loyalty accompanied by a higher level of quality does not always equate to customer retention unless supported by some innovations. According to management guru Peter Drucker, there are only two sources for growth: marketing and innovation [Drucker, 2008]. Innovation is often viewed as the Holy Grail of business and the primary driver for growth. Innovation forces companies to adapt to an ever-changing environment and to be able to take advantage of opportunities as they arise.

Companies are also aware that their competitors will eventually come onto the market with new products and services that will make some existing products and services obsolete, causing the competitive environment to change. Continuous innovation is needed, regardless of current economic conditions, to provide a firm with a sustainable competitive advantage and to differentiate themselves from their competitors.

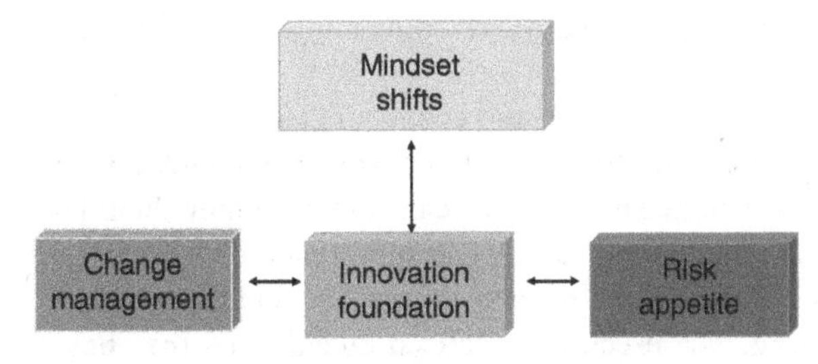

Figure 5.2 Innovation Foundation

The more competitive the business environment, the greater the investment needed for successful innovation. Companies with limited resources can take on strategic business partners and focus on co-creation. However, assumptions must be made as to whether the partners have assigned workers with the required skills.

For years, project management and innovation management were treated as separate disciplines. As indicated in Figure 5.2, innovation success requires building a foundation that has multiple ingredients, risk appetite, change management, and a mindset shift.

Innovation requires an acceptance of possibly significant risk, fostering of a creative mindset, and collaboration across organizational boundaries. Innovation management, in its purest form, is a combination of the management of innovation processes and change management. It refers to products, services, business processes, and accompanying transformational needs whereby the organization must change the way they conduct their business. It requires a different mindset than the linear thinking model that has been used consistently in traditional project management practices.

Project management practices generally follow the processes and domain areas identified in the PMIs ***PMBOK® Guide***. But now, companies are realizing that innovation strategy is implemented through projects. Simply stated, we are managing our business as though it is a series of projects. Project management has become the delivery system for innovation, and staffing is a critical component.

Today's project managers are seen more as managing part of a business than managing just a project. Project managers are now treated as market problem-solvers and expected to be involved in business decisions as well as project decisions. End-to-end project management is now coming of age. In the past, project managers were actively involved mainly in just project execution with the responsibility of providing a deliverable or an outcome. Today, with end-to-end project management, the project manager is actively involved in all life-cycle phases, including idea generation and product commercialization.

For decades, most project managers were trained in traditional project management practices and were ill-equipped to manage innovation projects. Project management and innovation management are now being integrated into a single profession, namely IPM.

Several years ago, a Fortune 500 company hired consultants from a prestigious organization to analyze their business strategy and to make recommendations as to where the firm should be positioned in 5 and 10 years and what they should be doing strategically. After the consultants left, the executives met to discuss what they had learned. The conclusion was that the consultants had told them "What" to do, but not "How" to do it. The executives realized quickly that the "how" would require superior project management capabilities, especially for innovation. Assigning the correct workforce was critical. The marriage between business strategy, innovation, and project management was now clear in their minds.

Organizations need the ability to manage a multitude of innovation projects concurrently to be successful, and therefore IPM is being supported by corporate-level portfolio management practices. IPM cannot guarantee that all projects will be successful, but it can improve the chances of success and provide much-needed guidance on when to "pull the plug," reassign resources, and minimize losses.

Types of Innovation

According to Webster's Dictionary, innovation can be defined simply as a "new idea, device, or method." However, innovation is also viewed as the application of better solutions. This is accomplished through more effective products, processes, services, technologies, or business models that are readily available to satisfy market needs.

From an organizational perspective, innovation may be linked to positive changes in efficiency, productivity, quality, competitiveness, and market share. Research findings highlight the importance of the firm's organizational culture in enabling organizations to translate innovative activities into tangible performance improvements [Salge & Vera, 2012]. As part of workforce planning, organizations can also improve profits and performance by providing workers with adequate resources and tools to innovate, in addition to supporting the employees' core job tasks. The tools that workers must use must also be considered in staffing activities.

Companies have been struggling with ways to classify the different forms of innovation. This has a serious impact on staffing. In one of the early studies on innovation, Marquis [1969] differentiated between incremental and radical innovation. Incremental innovation is a slight change to an existing product, whereas radical innovation is a change based upon a completely new idea. Incremental innovation focuses on existing markets and enhancements of existing products and services as well as refinements to production and delivery services [Danneels, 2002] and [Jansen et al., 2006]. Radical innovation focuses on the development of new technologies usually targeted for new markets, which adds a great deal of uncertainty and risk [O'Connor & Rice, 2013] and [Garcia & Calantone, 2002].

Exhibit 5.1 shows some typical categories for product development innovation. As we go from add-ons to complex systems, ambiguity and complexity will usually increase. Project managers may find it difficult to define requirements, understand changes in the marketplace, estimate time and cost, perform risk management, and deal with extensive meddling from stakeholders. Project managers may not be qualified to manage each type of product innovation project, and this does not include other company innovation projects such as service-related projects, new processes, and transformational projects. Effective workforce planning requires a good understanding of what type of innovation knowledge each worker might possess.

Some projects can begin as incremental innovation and then expand into radical innovation. This could require a change in team members. Project teams responsible for incremental innovation respond to changes in market conditions, whereas radical innovation teams perform in a more proactive manner. Radical innovation teams are multidisciplinary and self-managed. They must have open communication channels to share ideas and solve problems rapidly.

For innovation to occur on a repetitive basis and for a firm to retain its competitive advantage, the organization must create and nurture an environment conducive for innovation. Executives and managers need to break away from traditional ways of thinking and use change to their advantage. There is a great deal of risk, but with it comes greater opportunities. Innovations may force companies to downsize and re-engineer their operations to remain competitive. The impact that innovation can have on the way that a firm runs its business is often referred to as a disruptive innovation. We must remember that many process innovations result in disruptive changes rather than sales.

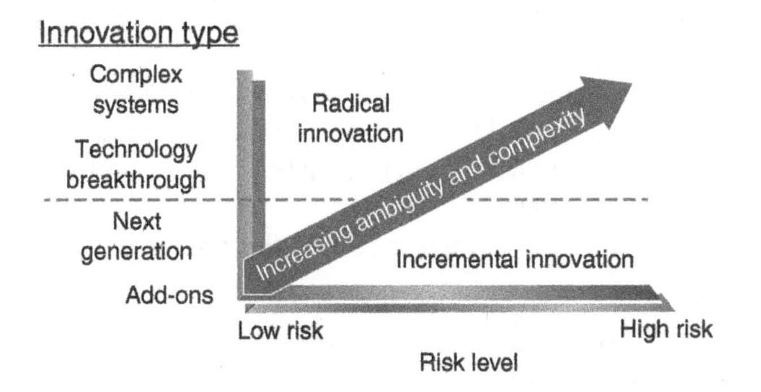

Exhibit 5.1 Typical Types of Innovation for Products

Wheelwright & Clark [1992], were one of the first to identify how diverse types of innovation projects can impact the way a firm is managed and that organizational change may occur frequently to maintain an innovation environment. The authors argue that companies are challenged with linking innovation projects to the company's strategy and that, although each project may have its own business strategy, there must still exist a linkage to the firm's overall business strategy. In an earlier work [Abernathy & Clark, 1985], projects were classified according to the firm's technical and marketing capabilities, which became the basis for identifying competence-enhancing vs. competence-destroying innovations. Therefore, workers assigned to projects requiring innovation should be familiar with the firm's business strategies.

The innovation environment in a firm can be defined as open or closed concerning idea generation and sources of information. Typical sources for ideas include:

- Customers, competitors, and suppliers
- Purchasing or licensing of technology
- Private inventors
- Academic institutions
- Government agencies and government-funded research
- Journals and other publications
- Technical fairs and trade fairs
- Open innovation fairs

Converting ideas to reality requires technology. There are four levels of technology for innovation:

- **Level I:** The technology exists within the company.
- **Level II:** The technology can be obtained from other sources within the country.
- **Level III:** The technology can be obtained from other sources outside of the country.
- **Level IV:** Technology must be researched outside of the country and brought back to the parent company.

The central idea behind open innovation is that, with knowledge distributed around the world, companies cannot afford to rely entirely on their own research. Intellectual property, including inventions and patents, can be bought or licensed from other companies [Chesbrough, 2003].

Open innovation offers several benefits to companies operating on a program of global collaboration:

- Reduced cost of conducting research and development
- Potential for improvement in development productivity

- Incorporation of customers early in the development process
- Increase in accuracy for market research and customer targeting
- Potential for synergism between internal and external innovations
- Potential for viral marketing [Schutte & Marais, 2010]

Implementing a model of open innovation is naturally associated with risks and challenges, including:

- Possibility of revealing information not intended for sharing
- Potential for the hosting organization to lose their competitive advantage by revealing intellectual property
- Increased complexity with controlling innovation and regulating how contributors affect a project
- Devising a means to properly identify and incorporate external innovation
- Realigning innovation strategies to extend beyond the firm to maximize the return from external innovation [West & Gallagher, 2006] and [Schutte & Marais, 2010]

Regardless of whether we use open or closed innovation, techniques must be established for the capturing and storage of information. Typical idea handling techniques include:

- Idea inventories
- Idea clearing houses
- Idea banks
- Screening or review teams

Innovation classifications can be made in the way that innovation data is collected and interpreted [Oslo Manual, 2005]. In this regard, innovations are represented as related project activities that can be classified as:

- Product/service innovations
- Process innovations
- Organizational innovations
- Marketing innovations

Innovation can also be classified by application. Keeley [2013], defines 10 categories of innovation by application:

- Profit model (How do we make money?)
- Networks (Do we have collaboration or partnerships with others?)
- Structure (Does our organizational structure help us and attract talent?)
- Process (Do we have knowledge, skills, and patents to sustain our processes?)
- Product performance (Do we have superior offerings?)

- Product system (Do we have products that are connected or distinct?)
- Service (Are customers happy with our service?)
- Channel (Do we have the right channels of distribution?)
- Brand (Do we have distinct brand identification?)
- Customer engagement (Are our products part of our customers' lives?)

There are many other classification systems for innovation. Companies with limited resources may adopt a "followership" innovation approach whereby the firm waits for the competition to develop new products and then tries to imitate it, produce it cheaper, and at a higher level of quality without having to recover the competition's innovation costs. Saren [1984] suggested classifying innovation projects according to five types: departmental-stage models, activity-stage models, decision-stage models, conversation process models, and response models. Pich et al. [2002] characterize projects based upon information available upfront to the project team:

- **Instructionalist project:** The information needed for innovation is available
- **Selectionist project:** Not enough information is available, and there is an elevated level of uncertainty
- **Learning project:** Susceptible to unforeseen events

Shenhar & Dvir [2004, 2007], identify innovation project categories as novelty, technology, complexity, and pace using their "Diamond of Innovation." These were some of the first articles that bridged innovation and project management. Another way to classify innovations is according to a complexity factor. One such approach identified five different dimensions of complexity: structural, uncertainty, dynamics, pace, and sociopolitical [Geraldi et al., 2011]. Other forms of innovation include:

- **Hidden innovation:** Performed under the radar screen and not reported with traditional metrics.
- **Discontinuous innovation:** Project direction must be changed in midstream because of changing conditions; rules of the game have changed quickly.
- **Disruptive innovation:** Causes some products to be removed from the market immediately and replaced with new products.
- **Crisis-driven innovation:** Must be done rapidly because of a marketplace crisis, such as designing a new package to prevent product tampering.

If any form of standardization is to be established for project managers, the starting point must be in the way that we classify innovation projects, which then determines how the project should be staffed. The number of different types of innovation makes it clear that innovation project managers and team members may need extensive training in IPM techniques, and the knowledge areas may be significantly different from how we trained them for traditional projects.

Co-Creation Innovation

Companies with limited resources often take on strategic partners as part of the innovation effort. The need for co-creation might be due to:

- In-house technical resources have insufficient knowledge/skills.
- In-house resources are committed to higher-priority projects.
- In-house talent exists, but work can be done externally for less money and in less time.

Regardless of in-house talent availability, co-creation allows the innovation project team members access to various supply chain members, including the end-of-the-line customers that are using existing products and those that might eventually purchase the innovation. Unlike traditional project teams that may interface mainly with in-house resources, the innovation project team becomes heavily involved with marketplace resources. Some of the significant benefits of co-creation with market users include:

- Supplanted Innovation
- Better alignment to the customer's needs and the customer's business model
- Maintain technical leadership and skills
- Investigate more new business opportunities
- Maintain a defensive posture to meet competition
- Balancing workloads and maintaining better asset allocation
- Maintaining customer goodwill
- Improvements to existing products
- Reduction in commercialization time
- Faster time to market
- Lower risk of failure
- Early identification of market reaction
- Better focus on value creation
- Reduction in innovation costs
- Make the company more competitive in the future and lower market entry barriers
- Repeatable elsewhere

The challenge with co-creation is identifying which people or actors in the marketplace should be part of the innovation team. Selection of co-creation partners must consider:

- **Who are the actors:** End users, extreme end users, suppliers, and distributors?
- Should we look at a mass crowd of users? (i.e. crowd-sorting)
- Do we select actors that are ordinary users or those possessing some proficiency in product development?
- Do we select ordinary/mainstream customers or advanced customers (lead users) that have superior knowledge and advanced needs?
- Should we select lead users in the idea generation stage and ordinary users in the testing stage?
- Does it make a difference in actor selection if we have incremental or radical innovation?
- Does it make a difference in actor selection as to what the innovation outcome is expected to be?
- Can ordinary customers think out of the box?
- What are the pros and cons of each group of actors?
- Can customer involvement vary based upon the life-cycle phase?
- How much freedom should we give the leading firm to design the architecture of participation and determine who makes decisions?
- How should we plan for situations where some partners are unwilling to share information if they are already participating in co-creation?

In traditional project management, the PM works with functional managers to staff the project team. With co-creation projects, such as in the case of pilot and co-pilot shown in Figure 5.3, the innovation project manager must work with senior management and consider critical issues such as:

- How do we recruit and find actors/customers?
- Will the actors function in just a support role or will it be joint innovation?
- Do we understand the cost and risks of co-creation activities?
- Who will own the intellectual property rights?
- Will there be a sharing or revenue and/or proprietary knowledge?
- Will any of the customers be unhappy and not agree to participate in the future?

There are situations that can occur, many of which are under the control of the innovation project manager, that can undermine co-creation efforts. Some of these include:

- Being close-minded about ideas of others
- Failing to keep creative people engaged and challenged
- Bringing on board the wrong people

Figure 5.3 Co-Creation Partnership

- Having too large a group
- Not listening to everyone
- Allowing time pressure to force you to make rapid decisions based upon partial information
- Criticizing ideas perhaps without justification
- Failing to focus on the lifetime value of the outcome

There are other factors that can destroy co-creation. They include:

- Working toward different goals and objectives
- Partner problems
- Working in a landlord-tenant relationship
- Cultural mismatch
- Personality clashes
- Working toward unrealistic expectations
- Lack of senior management support and commitment [Source: Adapted from Duysters et al., 1999]

TIP The future of innovation expects a workforce that is highly effective in applying co-creation as a muscle to drive creativity and to ensure buy-in.

Defining Innovation Success and Failure

Good as well as bad projects can fail. Defining traditional project success is easy; defining failure is difficult. We can consider success and failure in four categories:

- **Complete success:** The business goals were met, and the expected business value was obtained while meeting all the constraints on the project.
- **Partial success:** Some of the business goals were met and some value was obtained, but some of the constraints were not met.
- **Partial failure:** The only achievable business value was intellectual property that can be used in other innovation projects such as spin-offs.
- **Complete failure:** The project may have been canceled or simply not completed, and no business value, benefits, or intellectual property is indicated. Also, the outcome or deliverable may not have performed as planned [Kerzner, 2014, 13–14].

An executive in a Fortune 500 company complained that only about 20% of the innovation projects that were part of R&D were successful. The executive further blamed poor project management practices as the reason for the high percentage of "failures." The person who headed up the company's project management office (PMO) spoke up and said that "the other 80% of the projects were actually not failures but created intellectual property that the company will use on future innovation projects." The person also stated that "these projects showed where money should not be spent in the future so that mistakes would not be repeated."

The importance of these comments is that, even though an innovation project may appear as a failure by some because the business goal was not met, others may view it as a success because of spin-offs, lessons learned, identification of new opportunities, and an increase in knowledge. Failure can still provide some business value. When dealing with R&D and innovation, it is usually better to measure the success of the portfolio than the success or failure of a single innovation attempt.

The innovative capability of an organization rests in its ability to create knowledge. Not all innovation projects will be a technical success. The causes of innovation failure have been widely researched and can vary considerably even though the exact definition of failure is unclear. This will occur regardless of the knowledge of the assigned team members. Some causes will be external to the organization and outside its influence or control. Others will be internal and ultimately within the control of the organization. Internal causes of failure can be divided into causes associated with the cultural infrastructure, such as having a risk-averse culture, and causes associated with the innovation process itself.

What appears to be lacking today in both traditional and IPM is cancelation or exit criteria that provides guidance on when to pull the plug on a project. Without some exit criteria guidelines, poor projects can go on indefinitely, incurring large cost overruns even if economic conditions have changed. Any firm can design unique products that nobody will buy.

Exit criteria for a new product that has similar characteristics to the firm's existing products might include the following:

- The technology is not like that used in our other products and services.
- The product cannot be supported by our existing production facilities.
- The product cannot be sold by our existing sales force and is not a fit with our marketing and distribution channels.
- The product/services will not be purchased by our existing customer base.

TIP Proper definition of innovation success contributes to a higher likelihood that the estimating and proper allocation of the workforce is value-centered.

If the intent of the innovation is to break into a completely new area, then different exit criteria would be used. Typical exit criteria elements might then include:

- Insurmountable obstacles
- Inadequate know-how and/or lack of qualified resources
- legal/regulatory uncertainties
- Product liability risks
- Too small a market or market share for the product
- The product life cycle is too short
- Dependence on a limited customer base
- Unacceptable dependence on some suppliers and/or specialized raw materials
- Unwillingness to accept joint ventures and/or licensing agreements

Some companies prefer to define project suitability or success criteria during the selection and staffing process for innovation projects. Projects are judged and prioritized from the success criteria. The exit criteria are then created from not meeting the success criteria.

People that work on innovation projects must understand that their projects may have a greater likelihood of cancelation than other types of projects. Executives do not intentionally approve projects that have little chance for success. But because the risks and complexities of innovation projects may not be fully understood during the project selection process and the technical limitations of their innovation community may be unknown, a company must be willing to cancel projects that are later discovered to be poor projects.

Value: The Missing Link

The literature most commonly identifies three reasons for performing innovation; to produce new products or services for profitable growth, to produce profitable improvements to existing products and services, and to produce scientific knowledge, which can lead to new opportunities or problem-solving. But what about the creation of business value? The ultimate purpose of performing innovation activities should be the creation of long-term, sustainable shareholder value. Staffing innovation projects requires team members to understand the importance of value.

Value, whether business or shareholder, may be the most important driver in innovation management and can have a profound influence on how we define success and failure as well as how the project is staffed. Suitability and exit criteria must have components related to business value creation. However, it must be realized that financial value is just one form of value. Workforce planning can be impacted by the form of value that is expected.

Any company can make financial numbers look good for a month or even an entire year by sacrificing the company's future. Companies that want to be highly successful at innovation should resist selecting board members and even assigning certain workers to projects that focus mainly on financial numbers. From a strategic perspective, the primary goal for innovation should be to increase shareholder value over the long term rather than taking unnecessary risks and trying to maximize profitability in the short term.

There can be primary and secondary values created. As an example, a company creates a new product. This could be a primary value to the firm. If the company must modernize its production line to manufacture the product, then the modernization efforts could be a secondary value that could be applied to other products.

TIP As the maturity of defining project success increases, emphasis on value and balancing the views of key stakeholders will enhance the quality of workforce design.

The Innovation Environment

The innovation environment can be characterized by five words: ambiguity, complexity, uncertainty, risk, and crisis. While these words also apply to some degree to traditional project management practices, they may not have the severity of impact as in IPM [Pich et al., 2002].

Ambiguity is caused by unknown events. The more unknowns you have, the greater the ambiguity. As shown in Exhibit 5.2, there are unknowns in the innovation environment that are treated as known in traditional project management. There are several other differences that could have been listed in the exhibit. It is important to understand that the way we taught project team members in the past was by promoting the use of an enterprise project management methodology that had forms, guidelines, templates, and checks often designed to minimize the ambiguity on a project. These tools may not be applicable to innovation projects. Other tools will be necessary.

Exhibit 5.2 Some Differences Between Traditional and IPM Practices

Factor	Traditional Project Management	Innovation Project Management
Cost	Reasonably well known except for possible scope changes	Generally unknown
Time	Reasonably well known and may not be able to be changed	Generally unknown; cannot predict how long it will take to make a breakthrough. Innovators prefer very loose schedules so they can go off on tangents.
Scope	May be well defined in a statement of work and the business case	Generally defined through high-level goals and objectives rather than a detailed scope statement. Innovators prefer weak specifications for the freedom to be creative.
Work Breakdown Structure	May be able to create a highly detailed WBS (Work Breakdown Structure)	May have only high-level activities identified and must use rolling wave or progressive elaboration as the project continues
Resources Needed	Skill level of resources is generally predicable, and the resources may remain for the duration of the project	Skill level of the required resources may not be known until well into the project and may change based upon changes in the enterprise environmental factors
Metrics	Usually, the same performance metrics are used, such as time, cost and scope, and fixed for the duration of the project	Both business-related and performance metrics that can change over the life of the project must also be included
Methodology	Usually, an inflexible enterprise project management methodology	Need for a great deal of flexibility and use of innovation tools

Traditional project management focuses heavily on well-defined business cases and statements of work, whereas IPM relies upon goal setting. Tension can exist when setting IPM goals [Stetler & Magnusson, 2014]. There is no clear-cut path for identifying IPM goals. Some people argue that improper goal setting, or workers having personal agendas, can change the intended direction for an innovation project, whereas others prefer the need for some ambiguity with the argument that it creates space for innovative ideas, more fallback options are available, and the team may have an easier time converting ideas to reality.

Complexity deals with the number of components the project team must monitor and the relationship between components. Complexity also increases when the project team must interface with a large stakeholder base, all of whom may have their own ideas about the project. With a large stakeholder base, the team must deal with:

- Multiple stakeholders, each with a different culture and perhaps hidden agendas
- Political decisions become more important than project decisions
- Slow decision-making processes
- Conflicts among stakeholders
- Stakeholders that do not know their own role
- Frequent changes to the stakeholder base

Most innovation projects, which by nature are complex projects, generally have several components. The integration of these components requires an understanding of the relationships between the project, the company's business strategy, management practices, processes, and the organizational process assets [Gann & Salter, 2000] and [Hobday & Rush, 1999]. Although the complexity in innovation projects may be no different than the complexity in traditional projects, the impact it can have on risk management can be severe. Standard project management methodologies that often use linear thinking tend to evaluate risks on an individual risk basis without considering the human ramifications associated with each risk [Williams, 2017].

Ambiguity, complexity, and the accompanying risks are characterized by the amount of information available. If too little information is known, the payoff table for decision-making becomes ambiguous or there may be too many interacting parameters that create complexity. In both cases, risk management becomes complicated.

As complexity and ambiguity increase, so does the uncertainty. In traditional project management practices, we can assign a probability of occurrence to the uncertainty so that we can create a payoff table and perform risk management. But with innovation, there most likely will not be any historical data from which to assign probabilities to the uncertainty, thus increasing the risks and hampering risk mitigation efforts.

If we now include the impact on human factors, the situation can become even worse. When complexity, ambiguity, and uncertainty affect human factors, as they often do in innovation projects, risks can increase because the rational basis for decision-making may no longer exist. "In projects, bad things tend to happen in groups, not individually . . . Events that affect projects in many ways . . . tend to go together. Even when one of those things occurs individually, it tends to trigger a cascade of problematic effects" [Merrow, 2011]. The combination of risks, accompanied by management's actions and team reactions, can create vicious circles of disruption [Williams, 2017].

The fifth element characterizing the innovation environment is the need for crisis management. A crisis is an element of surprise that can occur unexpectedly and threaten the organization or the project. A crisis comes from a slowdown in the economy, recessions, wrong decisions, or unexpected events. In general, there are no contingency plans in place because of the number of different crises that can occur.

However, the innovation team must have a crisis mindset, which requires an ability to consider the worst-case scenario while simultaneously coming up with alternatives and solutions. Organizations must continuously use risk management and crisis management practices when there is a chance for significant unexpected events to occur.

The five elements characterizing the environment can have a serious impact on how a firm runs its business and the changes that may be needed for survival. As projects become more complex, the integration of components such as the firm's strategy, management practices, and organizational processes can and will change. We must understand the internal dynamics within the company [Gann & Salter, 2000] and [Hobday & Rush, 1999]. Once again, we see the importance of assigning workers that understand the business.

The Innovation Culture

The importance of an organization's and/or project's culture is often underestimated. Companies wishing for continuous innovation must create an organizational culture that allows people to freely contribute ideas. Workers must understand this when assigned to the project. In the same regard, creativity alone is not sufficient for achieving the goals of innovation; organizational initiative is a necessary condition for creativity to affect innovation. In a project environment, the PM must encourage people to bring forth ideas as well as alternative solutions and demonstrate his/her own commitment to the project. This also includes a visible willingness to accept risks. Executives are the architects of the corporate culture.

While the PM may be able to create a project culture, it must be supported by a similar corporate culture that also encourages ideas to flow freely, understands the strengths and weaknesses of the innovation personnel, and has confidence in their abilities. Companies can maximize the abilities of the innovation personnel by providing workers with a "line of sight" to the organization's strategic objectives and establishing processes to make this happen. When employees have a sense of awareness of the organization's direction and strategic goals, they make decisions based upon the greatest importance to the firm [Boswell, 2006] and [Crawford et al., 2006].

Effective decision-making is an important characteristic of the innovation culture. Decision-making should be structured and based upon evidence and facts rather than seat-of-the-pants guesses. However, because innovation isn't predictable, management must demonstrate a willingness to support tradeoffs and adjustments when necessary. Management must also make it clear to innovation teams that they have a willingness to cancel projects when certain criteria are not met. The cultural environment must be failure tolerant.

The meaning of "value" plays a critical role in establishing a culture. Good cultures create a mindset that value considerations are integrated in the way that project decisions are made. Some firms have value-driven cultures that focus on the delivery of business value rather than simply outcomes or deliverables from an innovation project.

There are however risks that need to be considered in any value-driven culture. They include possibly endless changes in requirements if the definition of value is not controlled; unnecessary scope creep while attempting to maximize value; having the definition of value made by different people over the project's life cycle; and stakeholders not believing the forecasted value.

The innovation culture thrives on the free flow of information. Some cultures struggle with information overload and have difficulty in evaluating the trustworthiness of the information. The more widely dispersed the project team is, the greater the need for effective communications and coordination. Effective communication and sharing of information create synergy. In the project culture, the team members must communicate with senior management as well as stakeholders. This must be done within a climate of trust.

TIP The project culture must be supported by an encouraging corporate culture where the flow of ideas and experimentation prevails.

Idea Generation

Some cultures spend more time collecting and analyzing ideas than using them. This happens if the ideas lack supporting data. As a result, not all projects are brought forth immediately. Most people know that the more information they discover to support their idea, the greater the likelihood that the idea can become a fully funded innovation project. One way to get supporting information, at least internally, is to create a bootlegged or stealth project. These projects are not recognized as "official" projects and do not have established budgets. The person with the idea tries to generate the supporting data while working on his/her other duties. If additional resources are needed, then the person with the idea must find people he/she knows and trusts to assist while keeping the project under the radar screen.

Bootlegged projects are done in secrecy because in most companies there is competition for funding and resources for other innovation projects. There may also be turf wars. Companies cannot fund or support all the ideas that come forth. Timing is everything. If the idea is released too early or if word leaked out about the idea, and without supporting data, there is a risk that the idea would be smothered by nay-sayers. These projects start in stealth mode because you can delay or postpone the moment that the clock starts ticking for your idea [Miller & Wedell-Wedellsborg, 2013].

When a project is done in secrecy, you may still need a sponsor to assist with getting resources and possibly some disguised funding. Generally, during the secrecy stage, you may be able to attract sponsors from middle management positions. Once the project is known, getting executive-level support may become essential because it gives you legitimacy, funding, and human resources. However, there is also a downside risk that your project and its team members are now in the spotlight, and careers may be at risk.

TIP Transparency and a culture of openness are critical elements for idea generation. The workforce of the future would play the role of idea ambassadors.

Understanding Reward Systems

When workers get assigned to a project, their first concern is, "What's in it for me?" They expect to be rewarded for the work they do. The fairness of the reward system can change behavior and affect risk-taking. Unfair reward systems can destroy an innovation culture.

Historically, reward systems were linked to cost-cutting efforts rather than innovations. This has now changed. There has been considerable research done on reward systems for product innovations [Jansen et al., 2006] and [Chen, 2015]. Reward systems generally follow two approaches. In a process-based reward system, which is often used in traditional project management practices, teams are rewarded based upon how well they follow internal policies, procedures, and expected behaviors to achieve the desired outcome. In an outcome-based reward system, teams are rewarded based upon the outcome of the project and the impact it may have on the bottom line of financial statements. There is significantly more pressure placed upon the workers in an outcome-based reward system.

In a pharmaceutical company, it may take up to 10 years and more than $1 billion to develop a new drug. The definition of a highly successful drug is usually expressed in dollars, such as generating more than $500 million yearly. Unfortunately, highly successful drugs may occur in less than 2% of the innovation projects. Therefore, someone could spend 40 years working on innovation projects and have no accomplishments that fall into this success category.

If compensation is tied to project outcomes, companies still seem to prefer to use existing well-established methods rather than seeking out new alternatives by trial and error. A better approach might be to tie the reward system to the risks that the project team must accept rather than entirely on the outcomes.

In radical innovation, workers are under a great deal of pressure to create innovative technologies for new markets using a highly uncertain development process that is accompanied by a multitude of risks. Individual motivation under these circumstances is critical [O'Connor & McDermott, 2004]. Employees must be trusted and given the freedom to experiment. However, boundaries must be set, and this can be done through goal setting [Pihlajamma, 2017].

Regardless of the reward system chosen, there are fears that workers may perceive. In IPM, there is the chance that people might resign if the reward system is unsuitable, if they do not receive recognition for their performance and ideas, if there is jealousy from the rest of the organization, and if the company has a low tolerance for failure. Reward systems must focus upon retention of talent and the ability to renew the firm's competencies. In traditional project management, the PMs generally have no responsibility for wage and salary administration. This may change in IPM.

TIP Reward systems could enable or stifle innovation. Innovation requires that the workforce is motivated to support working horizontally and focusing on discovery.

Resources Management

Executives tend to select projects, add them to the queue, and prioritize them with little concern if the organization has available and qualified personnel. Most executives do not know how much additional work they can take on without overburdening the labor force.

Balancing resource availability and demand requires open dialog. Innovation project team members for the project need to be brought on board early. Project managers need to participate in staffing activities and seek out qualified resources that support the idea for the project and are willing to work in a team environment. Some people may feel skeptical about the project. The PM must allay their fears and win over their trust. Project staffing requirements may dictate that the PM works closely with Human Resources for the duration of the project if people with new skills must be hired.

The importance of the Human Resources group is often hidden, but it does have an impact on creating a corporate image and reputation that promotes innovation. This is accomplished by attracting talented technical people, giving them the opportunity to be creative, and ultimately increasing the public's confidence in the value and quality of the innovations.

In fast-changing organizations, the link between strategy formulation and strategy execution is based upon the organization's understanding and use of dynamic capabilities. Dynamic capabilities theory concerns the development of strategies for senior managers of successful companies to adapt to radical discontinuous change. It requires reconfiguring assets to match a changing environment [O'Connor, 2008]. Organizations must have a firm grasp of the resources needed for competitive survival as well as the resources needed in the future for a competitive advantage. This can be accomplished using a talent pipeline that recognizes the competencies that are needed and their readiness to step in on short notice as backup talent. Specialized resources may also be needed because of deficiencies resulting from organizational change management.

There are shortcomings in resource management practices, as shown in Exhibit 5.3, which can prevent organizations from achieving their strategic goals and allow bad projects to survive. Executives may find it necessary to add resources to an apparently healthy project that has greater opportunities if successful. If the

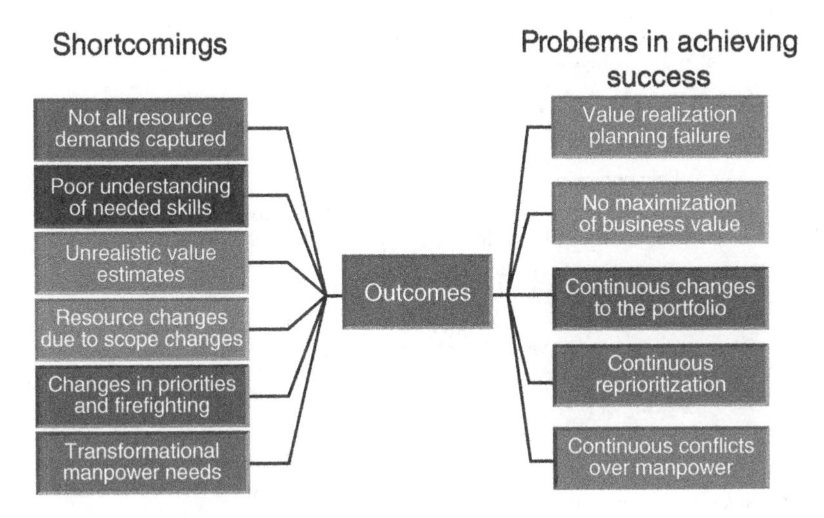

Exhibit 5.3 Shortcomings with Resource Identification

resources must be removed from another project, then the other project may have schedule delays and miss windows of opportunity. With a fixed workforce base, decisions must be made based upon the best interest of the entire portfolio rather than a single project.

Identifying the resources needed is part of the challenge. The other part is how the resources are allocated. Usually there is a priority system for resource assignment, as shown in Exhibit 5.4. Optimal resource capacity planning may be unrealistic. Some people believe that having organizational slack in resource assignments will increase the opportunities for creative behavior and contribute to a competitive advantage. There are three types of organizational slack:

- **Absorbed slack:** These are resources that are absorbed throughout the company but are recoverable later through efficiencies.
- **Unabsorbed slack:** These are resources that are available and can be assigned quickly to achieve a specific goal.
- **Potential slack:** This is the ability of the firm to obtain extra resources as needed through spending or raising funds [Murro et al., 2016].

Although we discuss organizational slack in terms of human resources, there can also be slack in physical resources and financial resources.

There are pros and cons for each category of organizational slack. In one company that prided itself on innovation, management created a culture whereby all workers were expected to spend at least 10% of their time on existing projects looking for ideas for new products for the firm. While this had the favorable effect of creating new products, it destroyed the budgets that project managers had for existing projects and was accompanied by significant cost overruns.

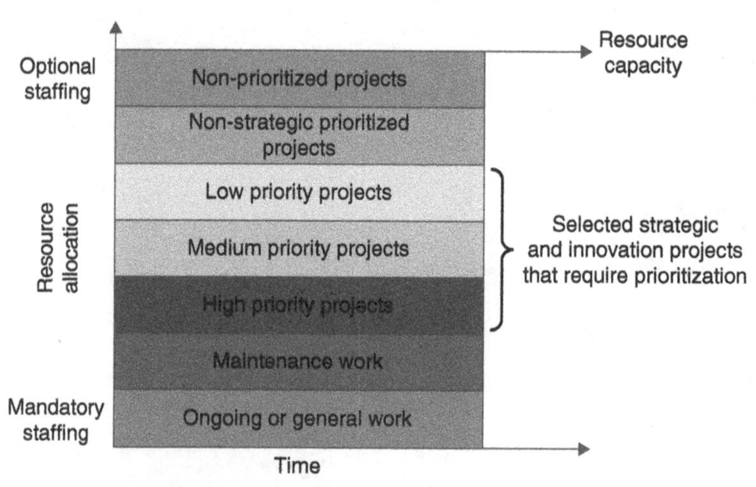

Exhibit 5.4 Resource Allocation

The Power of the Agile and Hybrid Approaches

In traditional project management, we rely heavily upon increasing the level of confidence that the governance models in place will help us achieve in controlling the destiny of the project. Over the years, agile practices were increasingly in use to focus on speed and value. This has always been the promise of project management, yet the language and way of working agile emphasizes, brought that forward more clearly to the focus of organizations and teams.

The workforce learned a great deal from the practice of agile and gradually expanded its use beyond the classical areas of technology and into areas like construction. This affected how work is done, the degree of uncertainty assumed, and ways of estimating the resources and efforts. As organizations matured in the practice of agile principles, it became more evident that it is not an either/or, but more of agile and. This meant that hybrid approaches seem to fit the most across organizations and industries.

As seen in Figure 5.4, agile practices emphasize simplicity, ease of transparent collaboration, and continual movement toward achieving value. The definition of done is a critical element of this practice. This focus is now expanded into a movement centered on the clarity of the definition of project success and maturing the workforce's views for how to focus their efforts invested in that. This will have a direct impact on the efficiency of work and thus result in estimates of quality enhancements.

Figure 5.4 Workforce Agile Approaches

The agile and hybrid approaches require a workforce that is adaptable and is capable to work across different models and ways of working. This is the future expectation that has been accelerated with the number of fast disruptions seen in this decade and that will only be magnified in the next decade. The next generation workforce needs to be able to learn insights from what works and what does not, as enabled by the rich data that captures patterns and allows for instilling the learning back and fast into the project work.

TIP The future workforce will utilize hybrid practices to shift fast toward value. Estimating project work will be enabled by the rich data gained from learning patterns.

Innovation Project Management Future Skills

In traditional project management, we rely heavily upon company policies and procedures. We may also have an enterprise project management methodology where the project manager simply instructs the team to fill in the boxes in the forms and checklists. With IPM, there will be different skills needed because the innovation project manager will be involved end-to-end in all the life-cycle phases shown in Exhibit 5.5. Perhaps the most important skill will be design thinking. Design thinking is a structured process for exploring ill-defined problems that were not clearly articulated, helping to solve ill-structured situations, and improving innovation outcomes. Design thinking can help resolve innovation challenges. Design thinking also mandates a close and trusting relationship between the team members and with the stakeholders throughout the life of the innovation. According to Mootee [2013], "Applied design thinking in business problem solving incorporates mental models, tools, processes, and techniques such as design, engineering, economics, the humanities, and the social sciences to identify, define, and address business challenges in strategic planning, product development, innovation, corporate social responsibility and beyond." Unfortunately, most of these topics are not covered in traditional project management training programs.

If the outcome of a project is to create customer value through innovation, then there is a need to bring design principles, methods, and tools into organizational

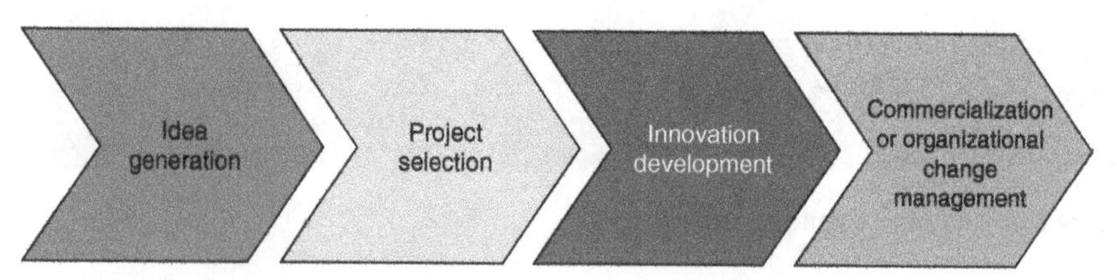

Exhibit 5.5 Typical Life-Cycle Phases

management and business strategy development [Brown, 2008]. "Design thinking and project management are both evolving rapidly as transformation factors and processes in firms and the economic landscape change. Both fields are anchored in a practice characterized by methods and tools, but they are moving beyond that operational perspective toward a strategic one" [Ben Mahmoud-Jouini et al., 2016].

There are more than 100 tools that can be used as part of design thinking [See Kumar, 2013]. Some common design thinking tools include:

- Storytelling (providing narrative info rather than dry facts)
- Storyboards (depicting the innovation needs through a story with artwork)
- Mind maps (connecting all the information)
- Context maps (uncover insights on user experience)
- Customer journey maps (stages customers go through using it)
- Stakeholder maps (visualizing stakeholder involvement)
- Personas (who are the users?)
- Metaphors (comparisons with something else)
- Prototyping (testing different ideas)
- Generative sessions (looking at stakeholder experience)

Brainstorming capability is another critical IPM skill, but some researchers disagree. The argument is that creativity usually precedes innovation, and the innovation project team may not be involved this early on. Creativity is where brainstorming takes place, and the role of innovation is to bring creative ideas to life. Innovation ideas can come from several sources, such as industry and market changes, demographic changes, new knowledge, and unexpected events. Innovation requires people to use both sides of their brain to take advantage of an opportunity. They must demonstrate diligence, persistence, and commitment regardless of their knowledge and ingenuity.

All ideas discussed during brainstorming sessions should be treated as intellectual property and recorded as part of a larger knowledge management system as shown in Exhibit 5.6. Idea management is knowledge management. Even if a company has idea screening criteria, all ideas should be recorded. What might appear as a bad idea today could end up as a great idea tomorrow. The drawback for a long time has been that innovation is a stand-alone process and not seen as part of any knowledge management system.

It is extremely difficult to get people to use a knowledge management system unless they can recognize the value of its use. Unfortunately, the only true value of a knowledge management system is the impact on the business in areas such as revenue generation, increased profits, customer satisfaction, and improved business operations [Hanley, 2014].

There are four activities that are part of knowledge management: knowledge creation, knowledge storage, knowledge transfer, and knowledge application. Organizational cultures can influence knowledge management practices by affecting employee behavior [Kayworth & Leidner, 2003].

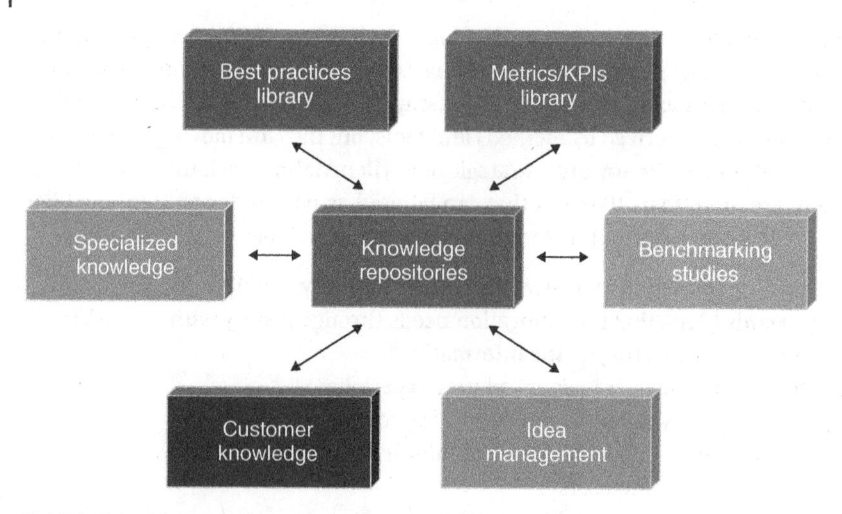

Exhibit 5.6 Typical Knowledge Management Components

There is a valid argument that everyone assigned to the innovation team should possess brainstorming skills. Walt Disney's Imagineering Division, which has the responsibility for designing theme parks around the world, is an example of how everyone throughout the life of the innovation project is expected to have brainstorming skills. At Disney, the term "Imagineering" is used, and it is defined as a combination of IMAGINation and enginEERING. Everyone in the Imagineering Division, from executives to janitors, calls themselves Imagineers and can participate in brainstorming sessions [Kerzner, 2017]. The culture in the Imagineering Division is totally supportive of brainstorming and innovation. Titles and silos are not considered during brainstorming efforts.

With projects requiring traditional project management practices, brainstorming may be measured in hours or days. The membership of the brainstorming group may be large or small and may include marketing personnel to help identify the specific need for a new product or enhancement to an existing product and technical personnel to state how long it will take and the approximate cost. Quite often, in traditional project management, a mistake is made whereby the innovation project managers or perspective team members may not be assigned and brought on board until after the brainstorming sessions are over, the project has been approved, added to the queue, and goals established.

TIP Future IPM workforce skills will build on the close alignment between the digital and human capabilities. A culture of emphasizing creativity of thinking will prevail.

Innovation Portfolio Management

There is a growth in the literature for an innovation portfolio PMO (IPPMO) dedicated just to projects requiring innovation. IPPMOs are critical for continuous innovations and can influence end-to-end IPM performance. There can be other PMOs dedicated to other functions, including strategic projects that may not require radical innovation of sorts. Unlike many other forms of portfolio management, innovation portfolio management is a complex decision-making process characterized by an elevated level of uncertainty. It deals with constantly changing information about opportunities internal as well as external to the firm [Meifort, 2015].

The IPPMO provides the necessary governance to link projects to strategic objectives. Innovation benefits the entire company, and therefore portfolio decision-making should be emphasized over silo decision-making. Analysis-paralysis situations by the workforce should be avoided. Finally, the gaps between the project team, various functional groups, governance personnel, and stakeholders should be reduced through effective communications.

Almost all projects require tradeoffs and, in most cases, the decisions about the tradeoffs are made by the project team. In innovation projects, the IPPMO may have a very active role and may be required to approve all tradeoffs as well as identify the need for tradeoffs because of the impact it may have on the business strategy and the need for change management. The IPPMO may have a better understanding of the changes in the marketplace and possess proprietary data related to strategic planning. Typical reasons for tradeoffs on innovation projects include:

- Loss of market for the product
- Major changes in the market for the product
- Loss of faith and enthusiasm by top management and/or project personnel
- The appearance of potentially insurmountable technical hurdles
- Organizational changes (i.e. new leadership with different agendas)
- Better technical approaches have been found, possibly with less risk
- Availability or loss of highly skilled labor
- Risks involving health, safety, environmental factors, and product liability

The IPPMO must insulate the innovation team from internal and external pressures. Some of the pressures include:

- Shortening development time at the expense of product liability
- Stockholder pressure for quick results
- Cost reduction
- Rushing into projects without a clear understanding of the need

Highly creative people thrive on recognition and want to show that their idea had merit and was achievable even if the market for their deliverable has changed. They do not like to be told to stop working or change their direction. As such, they may resist change and need to be monitored by the IPPMO if readjustments to the project are necessary.

Existing PMO literature focuses on the execution of projects that are reasonably well defined. Therefore, the roles and responsibilities of the membership in the traditional PMO can be reasonably defined. The IPPMO must serve as the bridge between innovation needs, business strategy, the organization's culture, and resource capabilities.

As such, the roles and responsibilities of the IPPMO membership are more complex. Perhaps the most significant role of the IPPMO is in the front end of innovation, where they must identify target markets, customer needs, value propositions, expected costs, and functionalities [Wheelwright & Clark, 1992] and [Bonner et al., 2002].

As stated previously, culture often plays a significant role in how companies create a portfolio of projects that includes innovation. Some cultures try to minimize risk and focus upon improvements or modifications to existing ideas, whereas more aggressive cultures pursue fundamentally innovative ideas with the goal of becoming a market leader rather than a follower. There may also be a national-level culture that influences project portfolio development [Unger et al., 2014].

Barreto [2010] identifies four responsibilities for the IPPMO: sensing opportunities and threats; making timely decisions; making market-oriented decisions; and changing the firm's resource base. Sicotte et al. [2014] add to the list topics such as intrapreneurship, proactive adaptability, strategic renewal, and value chain and technical leadership.

An important item that is frequently not discussed in the literature is the IPPMO's responsibility for nondisclosure, secrecy, and confidentiality agreements. This affects IPM more so than traditional project management. The IPPMO, working with top management, must develop a policy on how to handle the transfer of confidential information regarding innovation and technological developments to outside sources, including stakeholders.

Some of the significant differences from the traditional PMO that the IPPMO must perform include:

- Setting up boundaries related to the strategy so that reasonable goals can be established for the projects
- Deal with constantly changing information and opportunities
- Monitor the enterprise environmental factors
- Make sure that innovation specialists are assigned to the IPPMO to support opportunity-seeking behavior

- Support dynamic capabilities by determining if the organization must gain or release resources to match or create a change in the marketplace
- Look for ways to renew and vitalize the firm's competencies
- Balance the tension for resources between the ongoing business needs and the staffing for innovation projects
- Monitor the slack in the firm's resources because there is an associated cost, and too much slack may allow bad projects to survive
- Understand that resource allocation decisions are challenging because not all contingencies are known, and estimates and economic conditions are uncertain
- Monitor the performance of the projects to avoid design drifts

Critical success factors (CSFs) can be identified for effective innovation cultures, including IPPMO roles. Typical CSFs include:

- Senior management commitment acting through a culture that rewards innovative and entrepreneurial individuals
- Organizational structure and processes that support cross-functional teams and provide guidelines for their operations.
- Encouragement for new product ideas to be generated
- Providing venture teams with appropriate staffing, skills, resources, and training to be able to work and communicate effectively
- Promoting a tactical planning process for innovative projects that leads to shorter lines and earlier identification of pitfalls [Lester, 1998]

TIP The IPPMO plays a critical role in connecting the internal and external environments to value. This portfolio focus results in higher accuracy of future workforce estimating.

The Need for Innovation Metrics

Metrics are measurements. It is difficult, if not impossible, to manage what you do not measure. Mismeasurement is a root cause of mismanagement, especially when we measure just to collect data rather than measuring those items that make a difference. Simply put, we must measure what matters. Without good metrics, it is difficult to determine if you are heading in the right direction, and you may be unsure as to what you can and cannot do.

Perhaps the greatest difference between traditional project management and IPM is the need for more complex metrics. The workforce must understand the need for new metrics on these types of projects. For more than five decades, we have focused heavily upon Earned Value Measurement techniques stressing the measurements of just the time, cost, and scope because they were well understood and the easiest metrics to measure. Historically, when managing a project, things tended to move slowly, and we adopted an attitude of let's just wait and see what happens. Today, we must react faster than in the past.

Fortunately, measurement techniques have advanced to the point where we believe that we can measure anything. Projects now have both financial and nonfinancial metrics, and many of the nonfinancial metrics are regarded as intangible metrics. For decades, we shied away from intangibles. As stated by Bontis [1999], "The intangible value embedded in companies has been considered by many, defined by some, understood by few, and formally valued by practically no one." Today, we can measure intangible factors as well as tangible factors that impact project performance. The workforce must understand the need for both tangible and intangible information.

An intangible asset is non-monetary and without physical substance. Intangible assets can be the drivers of innovation [Kramer et al., 2011]. Intangibles may be hard to measure, but they are not immeasurable. In an excellent article, Ng et al. [2011] discuss various key intangible performance indicators (KIPIs) and how they can impact project performance measurements. The authors state, ". . . there are many diverse intangible performance drivers which impact organizational [and innovational] success such as leadership, management capabilities, credibility, innovation management, technology and research and development, intellectual property rights, workforce innovation, employee satisfaction, employee involvement and relations, customer service satisfaction, customer loyalty and alliance, market opportunities and network, communication, reputation and trust, brand values, identity, image, and commitment, HR practices, training and education, employee talent and caliber, organizational learning, renewal capability, culture and values, health and safety, quality of working conditions, society benefits, social and environmental, intangible assets and intellectual capital, knowledge management, strategy and strategic planning and corporate governance." Today, there are measurement techniques for these. In dynamic organizations, both KPIs and KIPIs are used to validate innovation performance.

With IPM, the workforce will need significantly more metrics, especially many of the above-mentioned intangibles, to track innovation and business applications. IPM metrics are a source of frustration for almost every firm. There are no standards set for innovation metrics. Companies may have a set of "core" metrics they use and then add in other metrics based upon the nature of the project. Metrics for disruptive innovation are probably the hardest to define. However, companies are beginning to develop models to measure innovation [Ivanova & Avasilcăia, 2014] and [Zizlavsky, 2016].

What is needed is a suite of metrics because no single metric will suffice. We must have metrics for each of the four life-cycle phases shown in Exhibit 2.4. This is where the IPPMO is of critical importance by asking the right questions, such as:

- Are we doing the right things?
- Are we doing the right things right?
- Are we doing enough of the right things?

These three questions can then be broken down in more detailed questions such as:

- Is project and portfolio value being created?
- What are the risks and are they being mitigated?
- When should the IPPMO intervene in any project decision-making?
- How will innovation performance affect future corporate strategy?
- Are the projects still aligned to strategic objectives?
- Do we need to perform resource re-optimization?
- Do we have any weak investments that need to be canceled or replaced?
- Must any of the projects be consolidated?
- Must any of the projects be accelerated or decelerated?
- Do we have the right mix of projects in the portfolio?
- Does the portfolio have to be rebalanced?

A common classification for metrics is product, service, and transformational metrics, which can then be broken down further into degrees such as incremental versus substantial transformation metrics [Lamont, 2015]. Metrics can be put into categories such as process measures, growth measures, and profitable growth measures [Linder, 2006]. Metrics can also be classified as activity metrics and impact metrics [Kirsner, 2015].

Some typical innovation metrics that the IPPMO might use include:

- Percent of projects in radical innovation
- Percent of projects in incremental innovation
- Percent of workers with the required competencies
- Revenue generated by new products over a given period
- Profit growth rate
- Profit and loss impact of innovation projects
- Profit and loss impact per customer
- Percent of projects in various life-cycle phases
- Project success rate
- Speed-to-market if applicable
- Number of innovation projects in the pipeline

- Number of patents
- Number of ideas generated
- Number of ideas killed (i.e. kill rate)
- Time to go from idea generation to project approval
- Process improvements such as in time-to-market

Financial metrics are usually easy to determine because the information is readily available. Most financial metrics generally do not capture all the value to the firm. Value-based or value-reflective metrics are more difficult to determine, but there are techniques available [Kerzner, 2013] and [Schnapper & Rollins, 2006]. Another challenge with value metrics is whether the value can be measured incrementally as the project progresses or only after the project's outcomes or deliverables have been completed [Kerzner, 2018].

All metrics have strengths and weaknesses. As an example, looking at the number of patents implies that the firm is creating technology for new products. But how did the number of patents affect the business? "The most important thing to remember is that . . . innovation is a means to an end, not the end itself, and therefore the most important metric is the contribution innovation and product development make to the business" [Lamont, 2015]. The timing of the measurement is also important. A common argument is that revenue-generating metrics such as sales and profits should be looked at over a predetermined time such as two or three years.

Some metrics are often ignored, such as those related to changes in culture and processes [Linder, 2006]. Another innovation metric that has been ignored until now is innovation leadership effectiveness [Ng et al. 2011]. Muller et al. [2005] have created a framework that includes innovation leadership metrics:

- **Resource view:** Allocation of resources
- **Capability:** Company competencies
- **Leadership:** Leadership support for innovation

Some firms start out with good intentions but either select the wrong metrics or measurement techniques, use misaligned metrics, or have a poor approach for innovation metric identification. Mistakes made often include:

- Too many metrics
- Too focused on outcomes
- Too infrequent usage
- Too focused on cost-cutting
- Too focused on the past [Kuczmarski, 2001]

In the future, innovation and business value metrics may become as common as time and cost metrics. However, the IPPMO must guard against "metric mania," which is the selection of too many metrics, most of which may lead to confusion and provide no value. When metric mania exists, people have difficulty discovering which metrics provide valuable information.

TIP With the increase in the number of value metrics required to pulse innovation progress, digitalization will help the future workforce make smarter metrics choices.

Extracting the Business Value

When an innovation project comes to an end, what you have is an outcome or a deliverable, accompanied by "perceived" value and benefits. Executives and governance personnel want to know if the promised benefits and value have been realized. Someone other than the innovation team may have to take ownership to harvesting the benefits and managing the transition period. Innovation is the means to the end, whereas sustainable business value is the end. As an example, you may have created an excellent product, but marketing and sales must now take the responsibility to harvest the benefits and value. You have created a new software package, but who will take the responsibility for the "go live" phase and train people in the use of the new software?

As seen in Exhibit 5.7, a company wishes to go from its current business value position to a new business value position. The company then authorizes funding for several innovation projects. When the projects are completed, the firm has outcomes or deliverables. Someone or some group must now take the responsibility for harvesting business value [Kerzner, 2018].

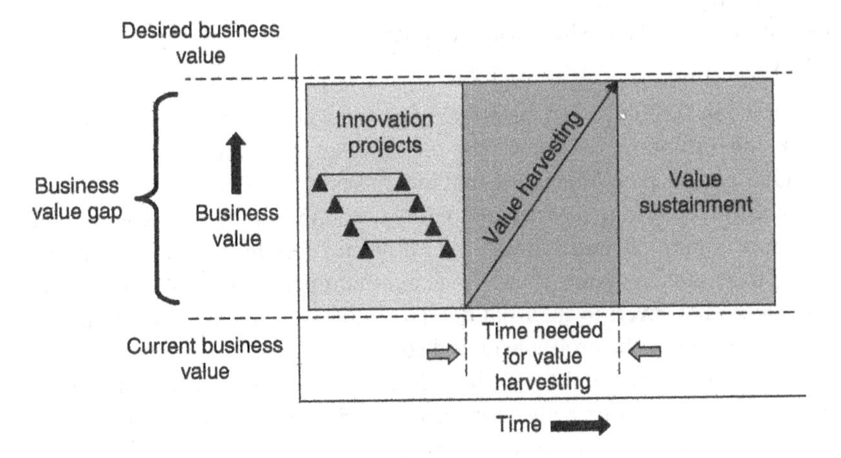

Exhibit 5.7 Value Extraction

The harvesting period can become quite long if organizational change management is necessary. People must be willing to be removed from their comfort zone. If organizational change is necessary, people responsible for benefits harvesting need to consider:

- Organizational restructuring
- New reward systems
- Changing skills requirements
- Records management
- System upgrades
- Industrial relations agreements

During value harvesting, people may have to change their habits, processes may change, and technology may change. Value extraction and full value realization may face resistance from managers, workers, customers, suppliers, and partners because they may be removed from their comfort zones. There is also an inherent fear that change will be accompanied by loss of promotion prospects, less authority and responsibility, and loss of respect from peers.

There can be costs associated with value harvesting if change management is needed. Typical costs may be:

- Hiring and training new recruits
- Changing the roles of existing personnel and providing training
- Relocating existing personnel
- Providing new or additional management support
- Updating computer systems
- Purchasing new software
- Creating new policies and procedures
- Renegotiating union contracts
- Developing new relationships with suppliers, distributors, partners, and joint ventures

Also shown in Exhibit 5.7 is a sustainment period where we try to extend the life of the business value we extracted. The real problem in many companies is not innovation that creates value but the inability to sustain it. Sustainment may be a necessity to increase long-term shareholder value, and many of the workforce team members may still be needed on the projects. If change management was necessary to extract the value, then the sustainment period is used to make sure that people do not revert to their old habits. Unlike traditional project management practices, where the project team is disbanded once the deliverables are produced, in sustainment, members of the IPM team may have a relationship with marketing and may still be actively involved in the project looking for product improvements, spinoffs, and future opportunities.

There is a significant need for effective IPM practices worldwide. The expectation is that training programs, and even certification programs, will be offered relatively soon focusing on IPM and emphasizing the differences that were discussed in this paper. The traditional view of project management as a "one size fits all" approach will not work.

Companies will find it necessary to adopt new forms of project management with more flexibility and trust given to the project managers. IPM will become a separate career path position, conceivably reporting to the top floor of the building because of the impact it can have on the company's strategic business plans.

The Value of Prompt Engineering in Fostering Innovation

There is a significant correlation between prompt engineering and the future workforce skills. Prompt engineering is found at the intersection of the human and the machine. This allows for the full power of GenAI and brings the project workforce to a position that puts forward their human strength to effective use and equips them with tools for higher project work efficiencies and thus enhances future project estimates.

Natural language is at the core of the increasing use of prompt engineering. This growing discipline is allowing us to be effective in the questions and dialog we have with the machine to get better insights that are valuable in driving the project journey forward. The exciting part about prompt engineering for project workforce is that the workforce that is trained well in portfolio and project management knows that asking the right questions is at the center of making proper investment choices leading to working on the right projects. Asking the right questions is also key to ensuring that the project work is done right. The future project workforce is ready to take on prompt engineering, and by learning this skill, a future shift in how work is done and where time is spent will continue to affect future project estimates.

This shift in the way of working, empowered by prompt engineering, allows the project workforce to be laser-focused on creating value. As imagined by Figure 5.5, prompt engineering is about developing a personal communication experience with the machine. It also highlights the need for holistic views of building that are energized by creativity and clarity of context.

Building the prompt engineering muscle for tomorrow's workforce is a change process that requires some trial and error, learning by practice, and a curious mindset. The conversational clarity and accuracy that are enhanced by specific guidance and stories will directly enhance the quality of outcomes received from the machine. The nature of future project tasks will also determine how much

Figure 5.5 Prompt Engineering and Creativity

prompt engineering will be of value. There will be project work that could be fully automated and some work that must be augmented and supported by humans.

This potential future state of project work will influence how project workforce estimates are developed and even the mix of the project workforce chosen for certain tasks. The growing value of power skills remains on an upward trajectory as we increase the reliance on digital solutions and artificial intelligence. This future state could be a future workforce that is more capable of addressing complex project problems. This workforce will be much more effective in addressing persona-specific needs, thus contributing to a higher joint perception of project success by many key stakeholders.

Among the most exciting aspects of prompt engineering is the ability to reach a future state where true project management capabilities could be exercised. This core shifts to better proactivity and having a risk-based way of working, coupled with being able to handle the organizational noises and the time-draining tasks well; all this contributes to a future project workforce that is more strategic and continues to better define project success around value and execute accordingly.

References

Abernathy, W. J. & Clark, K. B. (1985). Innovation: mapping the winds of creative destruction. *Research Policy*, 14 (1), 3–22.

Barreto, I. (2010). Dynamic capabilities: a review of past research and an agenda for the future. *Journal of Management*, 36 (1), 256–280.

Ben Mahmoud-Jouini, S., Midler, C. & Silberzahn, P. (2016). Contributions of design thinking to project management in an innovation context. *Project Management Journal*, 47 (2), 144–156. DOI: 10.1002/pmj.21577.

Bonner, J. M., Ruekert, R. W. & Walker Jr., O. C. (2002). Upper management control of new product development projects and project performance. *The Journal of Product Innovation Management*, 19, 233–245.

Bontis, N. (1999). Human capital valuation, working paper.

Boswell, W. (2006). Aligning employees with the organization's strategic objectives: out of 'line of sight', out of mind. *International Journal of Human Resource Management*, 17 (9), 1489–1511.

Brown, T. (2008). Design thinking. *Harvard Business Review*, 86 (6), 84.

Chen, Y. J. (2015). The role of reward systems in product innovations: an examination of new product development projects. *Project Management Journal*, 46 (3), 36–48.

Chesbrough, H. W. (2003). The era of open innovation. *MIT Sloan Management Review*, 44 (3), 35–41.

Crawford, L., Hobbs, B. & Turner, J. R. (2006). Aligning capability with strategy: categorizing projects to do the right projects and to do them right. *Project Management Journal*, 37 (2), 38–50.

Danneels, E. (2002). The dynamics of product innovation and firm competences. *Strategic Management Journal*, 23 (12), 1095–1121.

Drucker, P. F. (2008). *The essential drucker*. Reissue Edition. New York, NY: Harper Business.

Duysters, G., Kok, G. & Vaandrager, M. (1999). Crafting successful strategic technology partnerships. *R&D Management*, 29 (4), 343–351.

Gann, D. M. & Salter, A. J. (2000). Innovation in project-based, service-enhanced firms: the construction of complex products and systems. *Research Policy*, 29 (7), 955–972.

Garcia, R. & Calantone, R. (2002). A critical look at technological innovation typology and innovativeness terminology: a literature review. *Journal of Product Innovation Management*, 19 (2), 110–132.

Geraldi, J., Maylor, H. & Williams, T. (2011). Now, let's make it really complex (complicated): a systematic review of the complexities of projects. *International Journal of Operations and Production Management*, 31 (9), 966–990.

Hanley, S. (2014). Measure what matters: a practical approach to knowledge management metrics. *Business Information Review*, 31 (3), 154–159. DOI: 10.1177/0266382114551120.

Hobday, M. & Rush, H. (1999). Technology management in complex product systems (CoPS): ten questions answered. *International Journal of Technology Management*, 17 (6), 618–638.

Ivanov, C. I. & Avasilcăi, S. (2014). Performance measurement models: an analysis for measuring innovation processes performance. *Procedia – Social and Behavioral Sciences*, 124, 397–404.

Jansen, J. P., Van Den Bosch, F. A. & Volberda, H. W. (2006). Exploratory innovation, exploitative innovation, and performance: effects of organizational antecedents and environmental moderators. *Management Science*, 52 (11), 1661–1674.

Kayworth, T. & Leidner, D. (2003). Organizational culture as a knowledge resource, in Holsapple, C.W. (Ed.), *Handbook on Knowledge Management, Volume 1: Knowledge Matters*. Heidelberg: Springer-Verlag; 235–252.

Keeley, L. (2013). *Ten types of innovation*. Hoboken, NJ: John Wiley and Sons.

Kerzner, H. (2013). *Project management metrics, KPIs and dashboards*, 2E. Hoboken, NJ: John Wiley and Sons; 165–242.

Kerzner, H. (2014). *Project recovery: case studies and techniques for overcoming project failure*. Hoboken, NJ: John Wiley and Sons; 265–274.

Kerzner, H. (2017). *Project management case studies*, 5E. Hoboken, NJ: John Wiley and Sons; 255–286, 467–506, 583–654.

Kerzner, H. (2018). *Project management best practices; achieving global excellence*, 4E. Hoboken, NJ: John Wiley and Sons; Chapter 19.

Kirsner, S. (2015). What big companies get wrong about innovation metrics. *Harvard Business Review Digital Articles*, 2–5.

Kramer, J. P., Marinelli, E., Iammarino, S. & Diez, J. R. (2011). Intangible assets as drivers of innovation: empirical evidence on multinational enterprises in German and UK regional systems of innovation. *Technovation*, 31 (9), 447–458.

Kuczmarski, T. D. (2001). Five fatal flaws of innovation metrics. *Marketing Management*, 10 (1), 34–39.

Kumar, V. (2013). *101 design methods*. Hoboken, NJ: John Wiley and Sons.

Lamont, J. (2015). Innovation: What are the real metrics? *KM World*, 24 (8), 17–19.

Lester, D. H. (1998). Critical success factors for new product development. *Research-Technology Management*, 41 (1), 36–43.

Linder, J. C. (2006). Does innovation drive profitable growth? New metrics for a complete picture. *Journal of Business Strategy*, 27 (5), 38–44.

Marquis, D. G. (1969). The anatomy of successful innovations. *Innovation Newsletter*, 1 (7), 29–37.

Meifort, A. (2015). Innovation portfolio management: a synthesis and research agenda. *Creativity and Innovation Management*, 25 (2), 251–269. DOI: 10.1111/caim.12109.

Merrow, D. (2011). *Industrial megaprojects: concepts, strategies and practices for success*. Hoboken, NJ: John Wiley and Sons; 327.

Miller, P. & Wedell-Wedellsborg, T. (2013). The case for stealth innovation. *Harvard Business Review*, 91 (3), 91–97.

Mootee, I. (2013). *Design thinking for strategic innovation.* Hoboken, NJ: John Wiley and Sons; 60.

Muller, A., Välikangas, L. & Merlyn, P. (2005). Metrics for innovation: guidelines for developing a customized suite innovation metrics. *Strategy and Leadership*, 33 (1), 37–45.

Murro, E. V. B., Teixeira, G. B., Beuren, I. M., Scherer, L. M. & De Lima, G. A. S. F. (2016). Relationship between organizational slack and innovation in companies of BM&FBOVESPA. *Revista de Administração Mackenzie*, 17 (3), 132–157. DOI: 10.1590/1678-69712016.

Ng, H. S., Kee, D. M. H. & Brannan, M. (2011). *The role of key intangible performance indicators for organizational success.* Proceedings of the International Conference on Intellectual Capital, Knowledge, Management & Organizational Learning; 779–787.

O'Connor, G. C. (2008). Major innovation as a dynamic capability: a systems approach. *Journal of Product Innovation Management*, 25, 313–330.

O'Connor, G. C. & McDermott, C. M. (2004). The human side of radical innovation. *Journal of Engineering and Technology Management*, 21, 11–30.

O'Connor, G. C. & Rice, M. P. (2013). A comprehensive model of uncertainty associated with radical innovation. *Journal of Product Innovation Management*, 30, 2–18.

Oslo Manual (2005). *Oslo Manual – Guidelines for collecting and interpreting innovation data.* Paris: OECD and Eurostat.

Pich, M. T., Loch, C. H. & De Meyer, A. (2002). On uncertainty, ambiguity and complexity in project management. *Management Science*, 48 (8), 1008–1023.

Pihlajamma, M. (2017). Going the extra mile: managing individual motivation in radical innovation development. *Journal of Engineering and Technology Management*, 43, 48–66. DOI: 10.1016/j.jengtecman.2017.01.003.

Project Management Institute (2024). Talking to the machine. *Thought Leadership Series*, 4–9.

Salge, T. O. & Vera, A. (2012). Benefiting from public sector innovation: the moderating role of customer and learning orientation. *Public Administration Review*, 72 (4), 50–60.

Saren, M. A. (1984). A classification and review of models of the intra-firm innovation process. *R&D Management*, 4 (1), 11–24.

Schnapper, M. & Rollins, S. C. (2006). *Value-based metrics for improving results.* Ft. Lauderdale, FL: J. Ross Publishing.

Schutte, C. & Marais, S. (2010). *The development of open innovation models to assist the innovation process.* Roodepoort: University of Stellenbosch.

Shenhar, A. J. & Dvir, D. (2004). *Project management evolution: past history and future research directions.* PMI® Research Conference Proceedings, PMP, London.

Shenhar, A. J. & Dvir, D. (2007). *Reinventing project management: the diamond approach to successful growth and innovation*. Boston, MA: Harvard Business Press.

Sicotte, H., Drouin, N. & Delerue, H. (2014). Innovation portfolio management as a subset of dynamic capabilities: Measurement and impact on innovative performance. *Project Management Journal*, 45 (6), 58–72. DOI: 10.1002/pmj.

Stetler, K. L. & Magnusson, M. (2014). Exploring the tension between clarity and ambiguity in goal setting for innovation. *Creativity and Innovation Management*, 24 (2), 231–246.

Unger, B. N., Rank, J. & Gemünden, H. G. (2014). Corporate innovation culture and dimensions of project portfolio success: the moderating role of national culture. *Project Management Journal*, 45 (6), 38–57.

West, J. & Gallagher, S. (2006). Challenges of open innovation: the paradox of firm investment in open-source software. *R&D Management*, 36 (3), 319. DOI: 10.1111/j.1467-9310.2006.00436.x.

Wheelwright, S. C. & Clark, K. B. (1992). *Creating project plans to focus product development*. Cambridge, MA: Harvard Business School Publishing.

Williams, T. (2017). The nature of risk in complex projects. *Project Management Journal*, 48 (5), 55–66.

Zizlavsky, O. (2016). Framework of innovation management control system. *Journal of Global Business and Technology*, 12 (2), 10–27.

6

Designing the Future Workforce

LEARNING OBJECTIVES
• Understanding the importance of workforce engagement • Understanding effective workforce rewarding practices • Understanding staffing trends

Keywords *Understand workforce competency needs; Understand workforce recruitment practices; Understand effective workforce engagement practices; Understand the need for a workforce reward system; Organizational development*

Identifying Desired Team Competencies

The projects that we are working on today are larger and generally more complex than the projects we worked on in the past. Project managers are spending more time interfacing with the team and have expectations for the workers that are assigned to the project. Collaboration has become critical. The expectations include:

- They must know what they are supposed to do, preferably in terms of a product.
- They must have a clear understanding of their authority and its limits.
- They must know what their relationship with other people is.
- They should know where and when they are falling short.
- They must be made aware of what can and should be done to correct unsatisfactory results.
- They must feel that their superior has an interest in them as individuals.

Project Workforce Estimating: Best Practices for Project Managers, First Edition.
Harold Kerzner and Al Zeitoun.

Companion website: www.wiley.com/go/Kerzner_ProjWFE

- They must feel that their superior believes in them and is anxious for their success and progress.
- They must believe that they can benefit by working for you or on your project.
- They must believe that any promises made by you during recruitment will be kept. The functional organization will remember failures to keep promises long after your project terminates.

Project managers have limited authority in promising workers promotions, bonuses, future work assignments, and management-granted days off. Line managers have more control over these than project managers do. The future workforce is motivated by increasing levels of responsibility and clarity for linkages between the organizational strategy and their project work. This is something that the project manager can promise.

Perspectives on Project Management

Not all people working on projects have the same view of project management. This holds true in both project-driven and non-project-driven organizations. Some people view project management as a career development opportunity whereas others might view it as a non-promotable position and anxiously look forward to leaving the project.

As highlighted in Figure 6.1, these views could vary and some of the views team members might have include:

- People are often worried about what their next assignment will be after the project is completed. Unless this fear is overcome, they may create the need for additional work until another assignment is found.
- People can become confused when project policies and procedures are different from what they were accustomed to in their functional areas.
- People may worry that, if the project fails, it could result in a setback in their career.
- People know that the project organization is temporary and therefore may have more loyalty to the functional organization than the project organization.

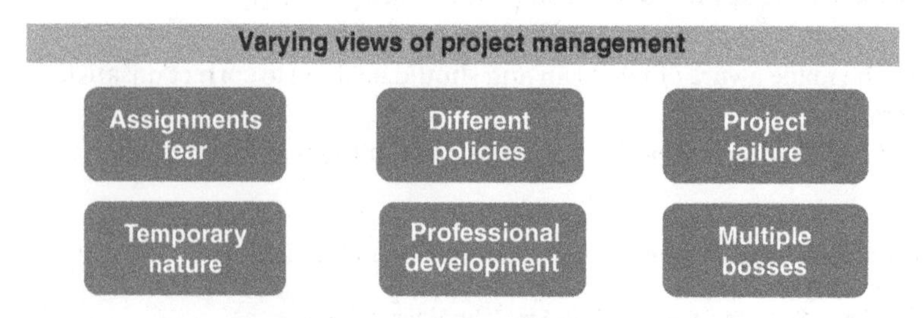

Figure 6.1 Project Management Perspectives

- People working on projects, especially long-term efforts, often worry that nobody is concerned about their professional development and career path opportunities.
- People who are used to functional organizations where you have one and only one person to report to are often frustrated when they discover that project management is multiple boss reporting.
- Frustrations caused by conflict are perceived more seriously in projects than in functional areas.

Strategies for Workforce Recruitment

There are certain recruitment policies that should be used when seeking project manpower. These policies hold true for both project managers who may have a say in who is assigned to the project and functional managers who may eventually make the final decision as to who is assigned.

Some of the recruitment policies might include:

- Unless some other condition is paramount, project recruiting policies should be as similar as possible to those normally used in the organization for assigning people to new jobs.
- Everyone should be given the same briefing about the project. This rule can be modified to permit different amounts of information to be given to different managerial levels, but at least everyone in the same general classification should get the same briefing. It should be complete and accurate. Special circumstances may exist when the information is company-sensitive or classified.
- Any commitments made to members of the team about treatment at the end of the project should be approved in advance by general management or functional management. No other commitments should be made.
- Every individual selected for a project should be told why he or she was chosen.
- A similar degree of freedom should be granted to all people, or at least all those within a given job category, in the matter of accepting or declining a project assignment.

Degrees of Permissiveness

There are several degrees of permissiveness that can be given to workers in regard to accepting or rejecting an assignment. The degrees of permissiveness can be given by the functional manager, the project manager, or both.

- The project is explained, and the individual is asked to accept the assignment. The individual is given complete freedom to decline, no questions asked.

- The individual is told by his/her functional manager that he/she will be assigned to the project. However, he/she is invited to bring forward any reservations he may have about joining. Any sensible reason will excuse him/her from the assignment. It may be a personal reason or related to a career preference.
- The individual is assigned to the project as he would be to any other work assignment. Only an emergency can excuse him from serving on the project team.

In an ideal situation, the project manager would like to have workers who volunteer to be part of the team. Usually volunteering breeds loyalty and commitment to the project.

TIP Of the three degrees of permissiveness highlighted, future workforce is more energized when they are given full decision-making autonomy.

Commitment and Expectation Management

Unless the project manager and the functional manager are the same person, project managers have virtually no responsibility for wage and salary administration. Yet some project managers try to motivate the team by making promises that cannot be kept. There are also situations where project managers make promises during staffing activities knowing full well that the promises cannot be kept. Both situations can have devastating results and may create significant interpersonal conflict.

Consider the following:

- Project managers cannot promise a functional employee a promotion. The project manager may be allowed to provide a recommendation to the functional manager who will make the final decision.
- Project managers cannot promise employees salary increases. This is a line function.
- Project managers may have money allocated in their budget for bonuses but may need the functional manager's approval.
- Project managers may have funding available for overtime on the project, but the final decision resides with the functional managers.
- Project managers cannot allow workers to perform work above their pay grade unless approved by the functional managers. This is especially true in collective bargaining situations or working with unions.
- Project managers cannot promise employees future work assignments. Project managers can request the worker to be assigned to future work but cannot promise the assignment.

Engaging High-Value Team Members

Project managers are at the mercy of the functional managers when it comes to the quality and skill level of the assigned functional employees. Project managers can request specific individuals with desired skills but the final decision about resources usually rests with the functional manager.

Some functional employees have outstanding technical strengths but simply do not work well in teams. These people work well by themselves and often resent supervision. It is the responsibility of the functional managers to inform the project managers about these people during the project staffing process.

Characteristics of a technical prima donna (a vain and temperamental person, a disagreeable person, or an unpleasant person)

- A desire to work alone
- A desire to work without close supervision
- When placed in charge of a team of people, will do all the work himself or herself and have little faith in the results of the team
- Must always validate other people's solutions before accepting them

Functional managers must provide some buffering to make it easier for the project managers to work with this type of person given the fact that the project manager may have no choice in the selection of resources.

Addressing Underperformance

Unless the project manager has worked with people previously, it may be impossible for the project manager to know whether the assigned resources will be lazy or underperforming. But once these facts are discovered, there are several ways the project managers can handle the situation:

- The worker may have been assigned a task that requires skills that he/she does not possess. It may be possible to reassign the worker to other tasks that match the worker's capabilities. It may be necessary to have the functional manager's permission to do this.
- It may be possible for the project manager to counsel the worker and identify the shortcomings of the worker. This assumes that the project manager clearly understands the job well enough to identify the shortcomings.
- The project manager can tell the worker that, if improvements in performance are not seen by a certain time, the project manager will provide an on-the-spot performance appraisal and send a copy to the worker's functional manager.
- The last resort is usually to approach the functional manager to whom the worker reports administratively and have the employee removed from the project team.

TIP Proper handling of performance topics requires a strong partnership between the project manager and the functional manager. Learning enhances future estimates.

Non-Financial Incentives for Motivation

Project managers generally have no direct responsibility for wage and salary administration. However, this responsibility depends upon the type of organization and organizational structure. If the project manager is also the functional manager there may be some responsibility regarding wages and salary. Therefore, project managers cannot provide direct cash awards to team members other than perhaps bonuses that have been included in the project's budget and are approved by the sponsor.

Project managers may be in a position to offer non-cash recognition or awards, and this can be done without the approval of the functional managers and/or the sponsor. However, it is often best for the functional manager to agree to this before it happens.

Typical non-cash awards include:

- Theater tickets
- Tickets to athletic events
- Certificates for fine dining
- Use of the company car
- Management-granted time off
- Plaques, newsletter articles, or other recognition methods
- Gifts from a catalog
- Paid vacation

These types of non-cash awards can be provided by the functional managers as well but are more commonly given out by the project managers.

Celebrating Achievements from Plaques to Public Acknowledgments

Sometimes the recognition associated with an accomplishment means more to the worker than receiving cash or something of monetary value. As an example, one company institutionalized a "Wall of Fame" that could be seen as the employees entered the company cafeteria. Whenever an employee did something outstanding for the company or the project, recognition was provided in the form of a plaque and mounted on the wall.

Another form of recognition is by publishing an article about the employee in the company's newsletter.

One insurance company provided certain employees with wooden tokens, like Boy Scout merit badges, for performance well done. The tokens were good for a free lunch or snack in the company cafeteria. What the company found was that the employees were performing exceptionally well trying to collect as many tokens as possible. The tokens were mounted in each employee's office for all to see. The motivation was to see how many you could collect.

The Role of Public Recognition

As mentioned previously, the recognition itself is often more important to the worker than the actual award or any dollar value that comes with the award.

Consider the following two examples:

- A company initiated a policy that any employee who becomes a Project Management Professional (PMP®)would receive a one-time bonus of $500. The company did not acknowledge publicly who became a PMP, but the next paycheck for the employee contained the bonus. The company then changed its recognition policy and once a week, in the cafeteria at lunch, an executive of the company would publicly recognize all of the people that became PMP®s during the past week. The bonus was still provided but more people were now taking the exam because of the public recognition.
- Employees that performed well were not only recognized publicly in this company but they were provided with elegant plaques ready for mounting. The employees mounted their plaques in their offices for everyone to see rather than bringing them home.

In both cases, the public recognition was more important to the workers than the award.

TIP Acknowledging the project workforce publicly could have a stronger energizing effect than just the pure financial awards.

Alternative Non-Monetary Rewards

As illustrated by Figure 6.2, there are other examples of non-monetary rewards.

- The company maintained a fleet of cars for the sales force and some of the executives. However, employees whose performance on projects was considered outstanding would receive use of the company car for a week or two. This was considered as a bonus by the employees that had to commute large distances each day.

Figure 6.2 Non-Monetary Rewards

- A company maintained a box at certain sporting events and certain theater events. Employees were given access to these seats for a job well done.
- A company completed a three-year project, and the customer was elated. Everyone in the company knew that the success of the project was due to one blue-collar union member who worked excessively overtime to make the project a success. Unable to reward him financially because of union policy, the company gave him the use of the company credit card for a week-long paid vacation for himself and his family. The union commended the president for recognizing the contributions of the employees toward the success of the company. Had the worker received say $6000 cash rather than use of the credit card, it is possible that a union grievance would have taken place.
- A company had a relationship with a very elegant restaurant where the company would entertain clients. Employees who performed well were given three free meals over a period of a month.

Public recognition does not necessarily require a formal reward and recognition event or an award. A simple "pat-on-the-back" or other expressions of thanks and appreciation can suffice. However, there are some people who fail to use reward and recognition properly or abuse the concept and its associated processes. Consider the following:

- A project manager believed that people should be told that they are doing a good job. This belief was based on a desire and a need to motivate the workers. The problem in this case was that the project manager was providing positive feedback even for an employee who was performing poorly to motivate the employee toward better performance. When these employees received a below-average performance review, they argued that the project manager indicated on several occasions that they were doing a great job.

- Rather than recognize everyone, a project manager believed that no one should be told they are doing a good job if they are merely doing the job they were expected to do. The project manager recognized the performance of only those people who went above and beyond what was expected of them. This led some people to believe that the project manager was displeased with their performance.

Timing and Patterns of Staffing Needs

Based on the size of the project, a staffing plan should be developed. There is often a mistaken belief that the intent of the staffing plan is to determine only what resources are needed and when. While this is true, the resource plan should also determine the ramp up and ramp down for the resources as shown in Exhibit 6.1.

The better the quality of the resources assigned by the functional managers, the greater the tendency that the functional managers will want these critical resources back as quickly as possible. Project managers tend to want to hold onto the resources, if possible, with the argument that something may go wrong. While this is true, the functional managers' argument is that these resources must service other projects and ongoing business activities.

Team members want to know when they will be released back to the functional areas. Some people believe that the longer they are removed from their functional manager, the less likely it is that they will be promoted. While this is not always the case, it is the way that some people think.

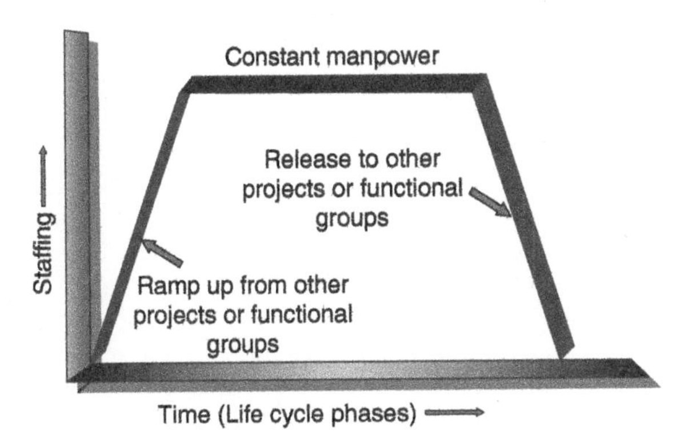

Exhibit 6.1 Typical Staffing Plan

The Role of Organizational Development

There is a growing maturity across industries associated with recognizing the importance of better organizational design and development. The design part has always been intriguing for project managers and the changing project workforce. Working horizontally across organizational boundaries, as necessitated by the nature of project work, is a critical work muscle to be developed. Organizational development with a focus on the speed toward value will continue to dominate future organizations' strategic agendas into the next decade.

Organizational development in a projectized future is a transformational topic. It is not about the past practices of just purely a wide net of topics that potentially will hit the mark for the workforce needs. It must be an intentional development path and must allow for the mix of people and fast-changing digital skills of the future. This development must be built on the chosen organization's design ingredients.

The following case study highlights some of the attributes to consider in organizational development. It sheds light on the increasing globalization in the future. It also makes it clear that estimating the future workforce will have to consider the increasing virtual work and that there will likely be no return to a pure office work format. Designing and developing the workforce to work in a hybrid work environment is future critical assumption.

CASE STUDY: THE RIVERSIDE SOFTWARE GROUP

The Riverside Software Group (RSG) was a medium-sized software company that specialized in software to support the Human Resources Departments of both large and small corporations. The RSG had been in business for more than thirty years and had an excellent reputation and an abundance of repeat business.

Since most of their work was global, they utilized virtual teams on almost all projects. The difficulty was in the creation of the virtual team and estimating the workforce cost of each global virtual team. Often, the project managers had limited knowledge of the capabilities of the employees around the world, and this made it difficult to establish a project team with the best available resources. What was needed was an inventory skills matrix for all employees.

The skills inventory software project was not that complex. The intent of the project was meaningful for future workforce estimating. Whenever RSG would complete a project for one of their worldwide clients, the entire project team would use the software to update their resumes including the new skills they developed, the tools or specialized processes they were now familiar with, and whatever additional information would be valuable to RSG in determining the best available personnel for the next similar project. The project team also had to identify in the software program the lessons that were learned on that project, the best practices that were captured, the metrics and key performance indicators that were used, and other such factors that could benefit the company in the future.

The Riverside Software Group saw this as an excellent opportunity, not only for RSG but for many of their clients as well. However, there were significant issues with the use of skills inventory software that needed to be overcome. Workers in different countries with the same job title and rank were at different salaries, and sometimes the salary differences were significant. RSG often has the resumes of key individuals in their proposals to improve their chances of winning the contract, but the key people would not be assigned to this contract after the contract has been awarded.

On some projects, the requirements were either unclear or needed to be changed after the go-ahead. This often required using higher salaried workers, thus lowering expected profit margins. Finally, the sales force often bids on contracts without a reasonable understanding of the workforce costs needed.

As highlighted in Figure 6.3, future workforce organizational development will be affected by several organization design dependencies. It is important that we develop the business, project, and digital acumen of the future workforce to accommodate as many of these dependencies as possible.

Strategic clarity is a cornerstone for development choices. Without that clarity, we would be investing in irrelevant skills and the design of the organization would not stick. The growing digital ways of working must be considered in the digital fluency that has to be embedded in the character of the future workforce. Realizing that structure changes should be done last is a good dependency as organizations must figure out what will be needed to meet its strategic intent first, before rushing into changing structures.

Building ownership is likely one of the most critical outcomes of future workforce development building an entrepreneurial mindset, understanding the organizational and project risk appetites, and strengthening the horizontal working patterns, are key contributors to future workforce ownership. Given the multiple generations in future workplaces, the ownership motivation will continue to vary and many of the reward and recognition topics addressed previously in the text could directly contribute to addressing this. Organizational development for ownership could benefit from

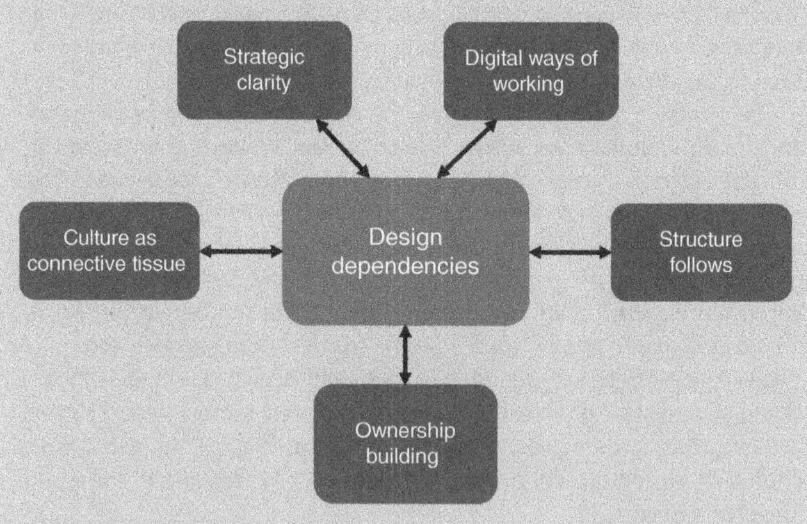

Figure 6.3 Organizational Dependencies

simulating situations of conflicts, changing environmental dynamics, and situations of uncertainty.

Last, but not least, culture is the connective tissue for both organizational design and development. Building a healthy culture for the future workforce is an imperative investment. Developing the future organizations' workforce to understand and live the values of the culture and effectively work in the specific nature of that culture is essential.

TIP Organizational development to address the future workforce skills is an essential element of project success. Fitting development approaches enhance estimates.

7

Advanced Topics in Workforce Planning

LEARNING OBJECTIVES
• Understand the need for planning for the unexpected • Understand workforce challenges • Understand the need for contingency planning

Keywords *Artificial intelligence; Life cycle costing; Management plans; Organizational structuring; Project funding*

Budget Allocation and Adjustment

Not all companies have formalized portfolio selection processes for determining which project to work on and in what order. In some companies, management may use rather unsophisticated techniques to select projects and then arbitrarily establish a sum of money for each project.

With this approach, there is often little basis for the size of the budget needed for a project. It may be just a guess. And to make matters worse, management might state that they refuse to hear bad news on the project, and they certainly do not want to be told that it cannot be done for this sum of money.

Once the funds are released for the portfolio of projects, line managers and executives begin fighting for their portion of the budget. The project manager, once assigned to a particular project, may have few choices with procurement costs, and the overhead rates are certainly not controllable. Therefore, the only flexibility is with the manpower: the number of people, the pay grades of the

Project Workforce Estimating: Best Practices for Project Managers, First Edition.
Harold Kerzner and Al Zeitoun.

Companion website: www.wiley.com/go/Kerzner_ProjWFE

workers, and the number of hours they will be given to do the work. This is unfortunate because the project manager must then select manpower to fit the budget rather than obtaining a budget that fits the manpower needed for the job.

If the amount of the funding is significantly less than the project needs, then project managers may be forced to accept lower-salaried workers or ask the team to perform the work in fewer hours. In any case, how the budget for the portfolio is partitioned may very well dictate which manpower will be assigned.

With advances in access to portfolio data, coupled with the maturing of enterprise software applications, some of these classic fights in portfolio decisions will change for the organizations of the future. Workforce planning could be enhanced with the potential objectivity provided by the data richness and the risk-based planning process that takes into consideration the many learnings and experiences organizations go through across their planning cycles. Even the classic approach of yearly planning is being disrupted and is becoming more fluid throughout the year.

Securing Additional Project Funding

Situations always occur on projects whereby additional funding may be required. Reasons for seeking out additional funding may include improper manpower estimating, technical difficulty requiring the use of more time or additional manpower, and escalations in salaries or overhead rates.

There are four primary sources for additional funding, and these are in the order that project managers may prefer to have them done:

- Customer-funded contractual changes: this implies that the customer requested or approved a scope change that requires additional man-hours or manpower.
- Management reserve: a management reserve is a sum of money withheld from the total allocated budget for management control purposes and some problems rather than designated for the accomplishment of a specific task.
- Undistributed budget: this is a sum of money, usually associated with contractual scope changes, for work that has been approved but not yet planned. It can also be part of next year's budget that has not yet been allocated toward specific work packages.
- Contract profitability: this is related to project-driven companies where, when mistakes are made, some of the project's expected profits are used to make corrections.

Global Workforce Estimation Challenges

Pricing out projects in foreign currency, especially if the host country has high inflation rates, can be troublesome. Managing a project that involves manpower from various countries also creates estimating complexities. In the United States, a senior engineer may be someone who has a certain number of years of experience, whereas in some emerging market nations, a senior engineer may be someone who has had two years of college education.

In some countries with high inflation rates, salary increases may occur monthly. Not all countries have their employees work eight hours a day. Some countries may have a 9- or 10-hour day for work, and others have 7-hour days. Some countries may have 20 or more paid holidays a year, and this can play havoc with manpower scheduling.

On projects that involve multiple countries, there is usually an agreed-upon currency or conversion rate that the client will pay for the work performed. Local currency will have to be converted into the agreed-upon currency, and this could be a problem if the local currency was devalued during the period of performance.

TIP Proper handling of global unique financial conditions is critical for sustaining the required workforce. This is more complex when combined with productivity variances.

Structuring Project Teams for Success

Large projects may be managed by a project or program office. The office includes the assistant or deputy project managers (APMs). There may be an APM for engineering, procurement, and manufacturing. The number of APMs depends on the size of the project.

The APMs are usually the lead people assigned from a functional area. For example, if you have 20 engineers working on your project, 1 of the 20 will serve as the lead engineer or APM for engineering. The project manager may not have much influence as to which employees are assigned to the project, but the project manager often has the authority to accept or reject the person assigned as the lead or APM. The project manager may end up working very closely each day with the APMs but not with the other assigned resources.

APMs must have good communication skills because they will be interfacing with the client and possibly the stakeholders. The APMs still report administratively to their functional areas. They have solid reporting vertically but dotted reporting horizontally. Some projects do allow the project manager to have wage and salary responsibility for the APMs, but this is not the norm.

Sometimes, there may be two APMs from the same functional area, such as engineering. Although their engineering responsibilities may be somewhat different, each serves as a backup for the other in case of an emergency.

As indicated by Figure 7.1, structuring project teams for success has dependencies on a multitude of topics. To name three of the ones highlighted across the

Figure 7.1 Team Structuring for Success

figure, there is an increasing focus on flexibility. The workplaces and the projects of the future are expecting a workforce that is adaptable and that is comfortable with change. The changing dynamics within and outside the project are increasing. Flexibility can be in role rotations, moving across teams, and in more expected multitasking. Competence is another one of these three that is becoming more difficult to define. In the classical workforce planning, the focus tended to be on the technical skills for the project at hand.

Now and into the future, the talent triangle keeps expanding, and areas like business acumen and leadership are moving to the front of the list due to their impact on project success. The third selected one from the figure is creativity. As addressed elsewhere in this work, the prediction by many practitioners and researchers is that the enhanced creativity and critical thinking will be such valuable assets in this highly integrated human-digital future workforce. Investing in building these qualities and predicting the impact they could have on project workforce estimating will allow the project manager to achieve higher estimation accuracy and make better workforce recommendations.

Incorporating Management Plan Data

On large projects, the client may request the right to approve people assigned to critical efforts because the client may be interfacing with these people daily.

This can include the project manager and all the APMs assigned to the project office. Clients that have worked with some of these people previously may request them on follow-on contracts.

People that work in the project or program office need not be full-time for the duration of the project. The engineering APM may be full-time during the design efforts and part-time during manufacturing. The manufacturing APM may be part-time during engineering activities and full-time during manufacturing.

Good workers may be in high demand on multiple projects. Clients know this and want to make sure that their project gets adequate attention. Clients may request that, during competitive bidding, the contractor identify what percentage of the critical worker's time will be spent on this project. The client may also request updates before approving the project plan. Some contracts have "key person" clauses that state that certain people must spend a given percentage of their time on this project, and the client has the right to cancel the project if this does not happen. This generally appears in the management section of a proposal that is part of competitive bidding.

TIP In order for the management plan to work well, the project manager should invest in relationship and trust building with the client. This will help mitigate changes risks.

Developing Contingency Plans for Workforce Management

We all run the risk that something unforeseen may happen to one or more of our critical resources. Sometimes resources are pulled from our project immediately to help put out fires elsewhere in the company. Sometimes people resign and leave the company immediately. Other times, people simply get sick or get hurt, and we end up with no qualified replacement.

Most well-managed companies develop succession plans for people in management slots. Each manager is expected to have someone in their organization ready to fill their position should they get transferred, promoted, or become ill. Years ago, large government programs overfunded the program management offices that were providing governance for the programs. People assigned to the program management offices were expected to serve as a backup for one or more program office workers should anything bad happen. The government recognized this as an over management cost and was willing to incur the costs.

Today, most project teams are running lean and mean. Functional organizations have sufficient resources and subject matter experts such that more than one person is qualified to fill a position. The learning curve for replacements in the functional ranks may be low. However, based upon the size of the project, it may be advisable to have one or more APMs assigned that can fill the shoes of the project manager in an emergency. The APMs may be part-time rather than full-time workers on the project. Obviously, the size, risk, and complexity of the project are the determining factors.

TIP Project managers own the succession planning for their role to sustain project team's continuity. These leaders learn that their job is to get themselves out of a job.

The Benefits of Co-Located Teams

There are both risks and rewards when using co-located teams rather than dispersed teams or virtual teams. The rewards are easy to see; the project manager has the entire team located in one area where monitoring and controlling of performance appears easier to do. But there are significant risks.

A few years ago, a government agency undertook a two-year project that required almost 120 government employees. Most of the employees were needed part-time rather than full-time. Unfortunately, the project manager wanted a co-located team for fear that the functional managers might assign the workers to other tasks, and they could be removed from the project. The project manager found a government building that had two floors that were vacant. All the employees were physically removed from their line managers' organizations and housed on the two floors where they would be under direct supervision of the project manager.

The team members were still attached administratively to their functional managers because the project manager did not have any responsibility for wage and salary administration. Also, when the project is completed, the workers would return to their previous functional areas.

While this appeared to be the right approach for the project manager, it had a detrimental effect on the workers' career development opportunities. During promotion cycles, the functional managers promoted the workers that were directly under their control and visible to them daily. The long-term co-located team discovered that they had limited promotion opportunities.

TIP With the increase in projectized organizations, it is critical that the project manager makes co-location decisions that support the proper career growth for the workforce.

Project Life Cycle Costing Approaches

Life cycle costing is the acquisition and ownership of a deliverable over its full or useful life. When planning a project, such as the construction of a new manufacturing plant, it is easy to say that we wish to lower our construction costs and therefore save money by utilizing less manpower. But the money you save now could be negligible compared to the money you might lose in the future.

Let's assume that the life cycle costing of a new manufacturing plant includes planning, construction, and operations and support of the facility once completed. The operations and support cost of the new plant could be as much as 80% of the total life cycle cost. You could probably save $2 million in the planning and construction phases. But what if the design of the plant is not optimal and the operations and support costs increase by $2 million each year above the expected costs? Your desire to save $2 million could now cost you $80 million over the life expectancy of the plant.

The purpose of life cycle costing is to minimize the total life cycle cost of the deliverable without any sacrifice in quality or performance. It is in the early life cycle phases where the critical decisions are made that lower the overall cost. A shortage of manpower in these phases can be costly later.

This is a critical capability for the future workforce. Being able to think end-2-end and not in a siloed fashion when it comes to lifecycle costing also has a direct relationship to the project value achievement. Developing this view opens the possibility for better ownership of value at the right stages of the lifecycle. This requires that the project manager and the project workforce operate more strategically in how they assess costs and how to prioritize any shifts across the lifecycle so that an opportunity in one part of the project does not end up threatening other project parts.

Techniques for Workforce Leveling

One of the challenges that project managers face is how to deal with the way that manpower comes and goes on a project. In Exhibit 7.1, the dotted line represents the peaks and valleys in manpower. Project managers would prefer to have the same faces on a project all the way through, but this could be too costly for the project.

To resolve this problem, the project manager usually resorts to workforce smoothing or leveling. Workforce smoothing is an attempt to eliminate the manpower peaks and valleys by smoothing out the period-to-period manpower requirements. The ideal situation is to do this without changing the end date. However, the end date usually moves out, and additional costs are incurred.

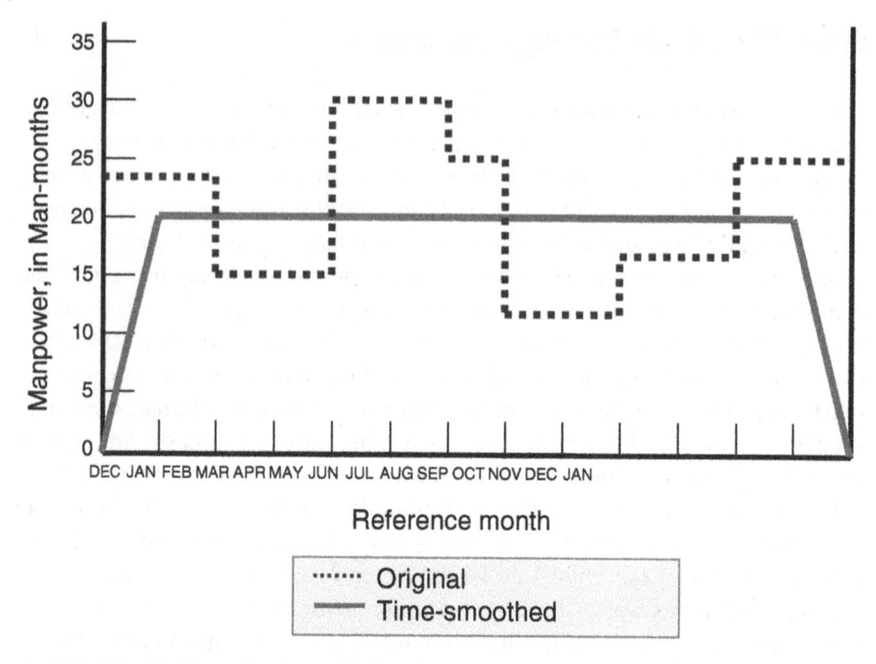

Exhibit 7.1 Workforce Peaks and Valleys

Without a workforce curve, peaks are troublesome for project managers because there is no guarantee that the same resources will appear at each peak. This could result in a loss of learning, and new people may require additional time to catch up to their colleagues.

TIP One of the biggest advantages of portfolio management software is the ability to have a transparent holistic view of project workforce, thus balancing the workforce use.

Advanced Workforce Leveling Strategies

If project managers bring manpower on board too early, there will be an added cost. In Exhibit 7.2, we see the planned workforce smoothing curve. If the project manager wants to save some money, then the project manager may need to change the rate at which the project is staffed and de-staffed. The ramp-up and ramp-down must then be more gradual, as can be seen in the modified time-smoothed line.

In Exhibit 7.2, the planned time-smoothed curve required a maximum of 20 people, and the modified time-smoothed curve required a maximum of only

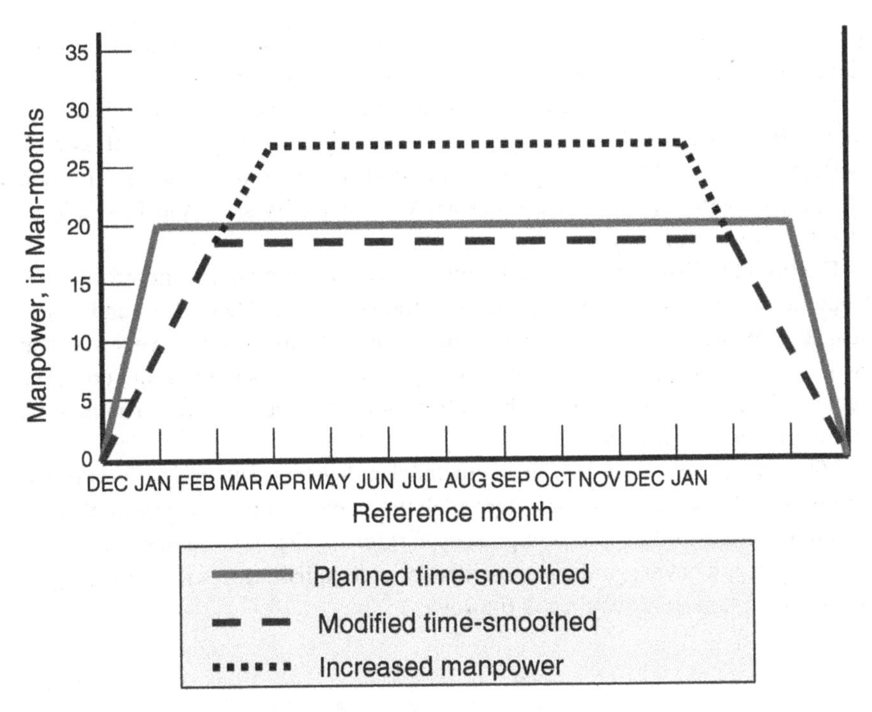

Exhibit 7.2 Adjusted Workforce Smoothing

19 people. But what if we have a shortage of people to do all the necessary work, especially the work on the critical path? The solution may be to use the increased manpower curve and have 26 people assigned at the maximum point.

Manpower smoothing works well if:

- The scheduled end date is allowed to slip
- Some cost increases are allowed to compensate for the smoothing
- Work packages, perhaps even some on the critical path, are allowed to move upstream or downstream to allow for a smooth manpower curve.

Maturing an Inventory of Skills and Competencies

Some companies have created databases for the inventory skills of the workers. Whenever a project is completed, the workers must update their inventory skills matrix by identifying all new skills that they developed on the project that was just completed. If, at the beginning of a project, the project manager is looking for people with a specific skill, he/she can simply input that desired skill into the database and get a complete listing of all the workers with that skill.

The database can also be used to validate that the workers have the necessary skills. Whenever people are assigned that the project manager has not worked with previously, and they are assigned to positions that require critical skills, then the project manager can pull up that worker's inventory skills sheet to validate that the capability is there.

The intent of Figure 7.2 is to highlight the need to maintain the motivation and appetite to continue to grow the skills of the workforce. There is no end to the learning potential for the future workforce. The skills mix changes every day. The access to new skills is easier, and with the accessibility being feasible on every technology platform possible and in multiple formats that allow this to happen on one's time more consistently.

With all the advancements in competencies development, project workforce seems to still lack consistent ways by which the categorizing of the groups of skills could be standardized across a large-size workforce. Also, linking these skills to a career path and having guidance provided for accomplishing this mapping by professional organizations could still improve.

Figure 7.2 Maturing Competencies

The Next Potential of AI in Enabling Workforce Planning

A new era has begun. This is an era that is witnessing an increase in the utilization of digital in every aspect of our lives. In the project environment, this is offering a promise for the highest degree of efficiencies in planning and executing throughout the project lifecycle. Just as in the case of many of the revolutions that affected organizations and the ways of working over the years, this Generative AI (GenAI) impact is bound to be central to project work for years to come.

As seen in Figure 7.3, there needs to be a successful bridging between the human workforce and the digital universe that is fast expanding around today's projects. The upside for this bridging is massive. It does require an appetite for experimenting with cultures that support that, leadership that sees the business environment objectively and holistically, and ultimately project managers and project teams who are ready to adopt fast and scale faster. The future of work will witness a demand that our project estimates are done faster and that changing circumstances could easily be incorporated to revise these estimates.

Let's tackle six factors that will either contribute to scaling the potential of artificial intelligence (AI) or limit it. The project manager is expected to take the lead in understanding these factors and in being the coach for the future workforce as they further dive into the experimentation and use cases surrounding AI's potential. Some of the use cases are internal to the organizations as they change themselves and how they work; other use cases affect how organizations

Figure 7.3 The Human-Digital Bridge

imbed AI in their products and solutions and position that value to current and future customers. Each of these six factors can function as an enabler to scale the future use of AI in better equipping the workforce. These enablers break change resistance barriers that currently exist across organizations and project teams.

Like many other areas of project management, the question comes to the surface regarding ownership. Just like when we ask who should own project success, this ownership topic could be even more challenging in the case of AI. In an ideal state, AI should be owned by everyone since it is a strong enabler that could be used differently depending on who the stakeholder is and what impact is expected by the workforce from its use.

The first of the six factors is ***Culture***. Culture is a mix of many things, ranging from values to behaviors and ways of working. Culture reflects itself in the walking of the talk by leaders. The workforce of the future is expecting organizational cultures that are open, motivating, and rewarding to the generational mix in the room. Looking at the enabling elements of the culture for AI, the expectation is to have the culture as exemplified by the messages sent by executive leadership to encourage the access and the use of AI.

As shown in Figure 7.4, a strong commitment in the culture to be ready operationally and by building the right safe infrastructure is a must. Cultural readiness also includes the investment in being educated and trained in AI across the layers of the organization. Having a clear strategic agenda for the use of AI is critical. A clear strategic choice about all the possibilities for AI use needs to be obvious on the NorthStar of the leadership team and should cascade across the workforce. Having this clear commitment will make it easier for the workforce to achieve all the potential benefits and will directly contribute to higher efficiencies in the future and lead to more competitive estimates.

The second of these factors, in Figure 7.5, is the ***Human Skills*** building factor. This is probably one of the most critical workforce changes that must happen. Not only does the workforce need to build the strong bridging to technology discussed earlier, but also the workforce should be open itself to a new mindset. It is a digital growth mindset that recognizes the limitless possibilities that technology provides. It is also a mindset that is ready and willing to be reshaped by a different

Figure 7.4 The Culture Enabler

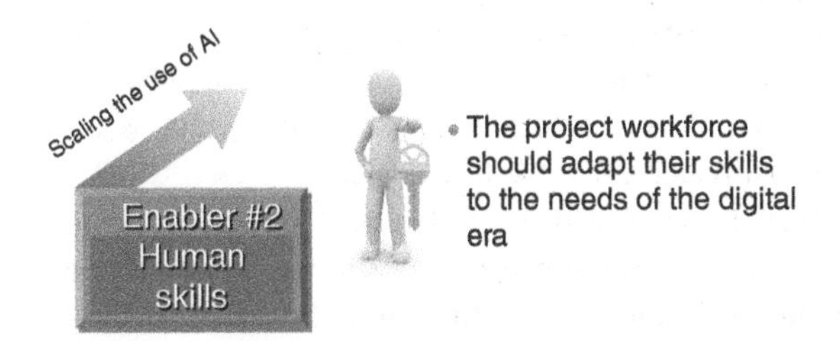

Figure 7.5 The Human Skills Enabler

mix of skills. The emphasis here is on the human skills. The future of work, although digital, is human-to-human. By experiencing higher use of technology and its potential, this creates a wider space for project workforce to pay attention to the human side of project work. When the workforce can think again for a change, it becomes more strategic.

It is time to become more empathetic and build that emotional intelligence edge. This is a great starting point. The mix of skills will have to include prompt engineering as previously discussed when the workforce is going to get better outcomes from the machine with the right strong context and examples used in the prompts. In addition, one of the most important skills will likely be around creativity and critical thinking. We have known for years that the effectiveness of solving the most complex problems faced in projects and programs hinges on this advanced thinking, yet the maturity has never been fully achieved given where project teams have been spending their time and the many other classic project distractions and firefighting.

Given the space created by AI, it is likely that the future workforce will be able to invest more in the Power Skills (human, leadership, and people skills). This will allow for more intentional interactions, enhanced diversity of ideas, and a better potential for integrating that with data toward the achievement of unique outcomes out of the projects and meeting the many innovations initiatives demands discussed earlier in this work. The future workforce could have more interesting project work and could enjoy seeing the outcomes of their work better than they have been able to do in the past.

The third factor is the ***Career promise***. It is assumed that with the enhanced use of AI, there is a higher likelihood that the nature of work for the project teams becomes more attractive. Most importantly, when it relates to careers, the value proposition of the project workforce could become clearer. This will provide a line of sight to where the potential of the project workforce lies in the organization and cross-teams could be.

The core here to the enhanced career potential is the shift of focus on value and the fast movement toward a new view of project success. With clarity that success is no longer achieved via the classic drivers and constraints of projects of the past, namely time, cost, scope, and a few others, and rather more about what the achieved value for the customers and users looks like and how that value is sustained, the longevity of the workforce career is likely stronger.

For the career promise enabler shown in Figure 7.6, AI could provide an opportunity for additional areas of specialization in project work. Data scientists, analysts, and others who can put technology, analytics, and the many insights garnered from the use of technology to better use will have an additional opportunity to create new forms of needed jobs in the future and an ability to combine these digital skills with the creativity and leadership qualities needed under the human skills reviewed earlier in this work.

The next factor is ***Trust Foundation***. Trust is normally a foundation for effective workforce interactions in projects and programs. It becomes more of an important currency in the future of work given the shifts necessary to become comfortable with digital future work. There is a level of trust that should be built for the wide use of AI. This means that we must design and implement responsible AI. This covers areas like proper use, cybersecurity measures implementation, and the importance of respecting the intellectual property of the organization. There is still a high degree of hesitation in expanding the use of AI related to this trust topic.

Like building high-performing teams on a foundation of trust, such as in the work done by Partick Lencioni, it is also critical to build the muscles of being able to have conflict and critical conversations on where GenAI is to be used and where it should not be. It is also needed that the executive leadership is very clear in its message about the impact of AI on the career promise topic addressed above. To have proper buy-in and commitment from the project workforce, this trust foundation is immensely important.

As illustrated in Figure 7.7, new leadership might be required in future organizations. Leaders should be able to exemplify their comfort with AI. They should also be able to communicate with a very clear voice that shows care for the workforce and responsibility for the organization. This builds on the qualities project

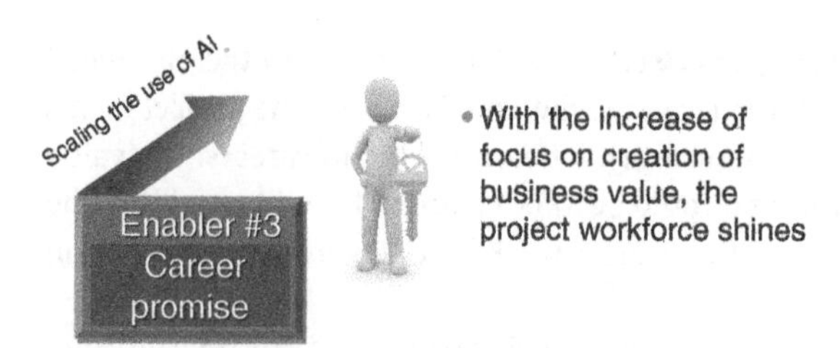

Figure 7.6 The Career Enabler

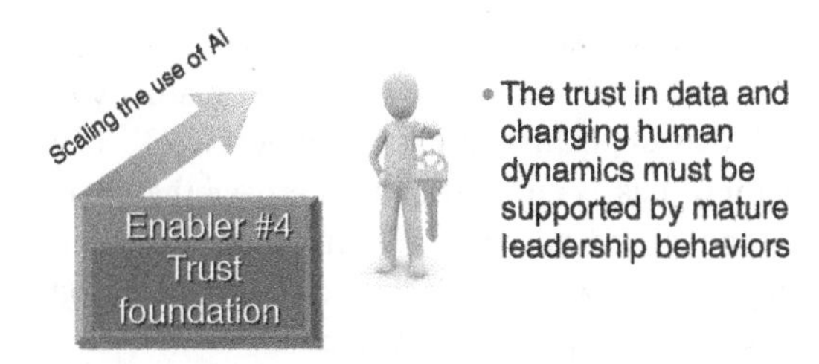

Figure 7.7 The Trust Foundation Enabler

teams have been striving to secure for decades in the type of leadership project teams should possess.

The fifth of these factors highlighted in Figure 7.8, is ***Risk Appetite***. With commitment to AI and its use, there is a higher degree of transparency across the workforce. This means that we can have access to sometimes uncomfortable data in a timely and it also means that we are able to operate more accurately and with excellence. This requires a workforce that is comfortable with change. The risk appetite and willingness to experiment need to be high.

To come up with meaningful use cases, the project workforce should be willing to take more risks. This requires the strong backing of the organization and its leadership. Such as discussed in the first factor, culture, a degree of safety in operating and openness exemplified by the leaders is a must. When this support is obvious, the project workforce will be able to take on more risks, have less fear of the consequences of AI, and be less worried about the impact of AI on the careers of the project team members.

In an ideal state, and when the room for taking risks is higher, the project workforce could be working on more meaningful tasks and worry less about biases in performance. Reporting and other factors that would have typically affected the workforce' ability to perform strategic work could be removed. To ensure that there will be less data bias, the point about new forms of leadership discussed above becomes crucial. Leaders in the future need to exemplify authenticity and possess a high aptitude for exercising the duty of care for their organizations.

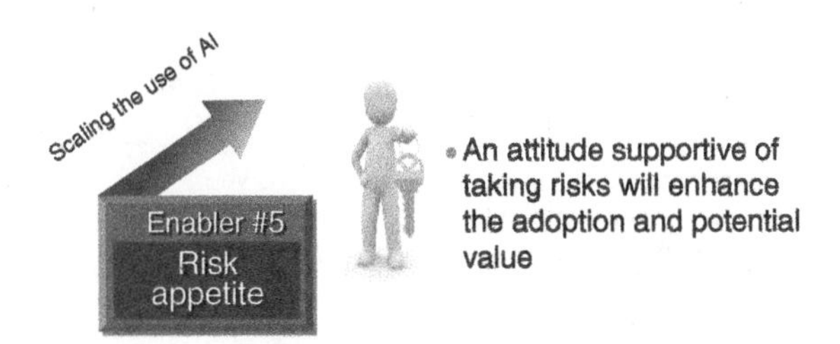

Figure 7.8 The Risk Appetite Enabler

The last of these factors is the ***Value-Based Decisions*** enabler. Decision-making in projects is one of these areas that shows the right signs of maturity in project work. For decisions to have the most positive impact on project work, they should be value-based. Although there has been a wide debate over the years about suitable project delivery methods and whether this enhances decision-making or not, the project manager should own the selecting of the right fitting delivery method that matches the needs of the project workforce.

Value-based decision-making, highlighted in Figure 7.9, is where scaling the potential of AI could be multiplied. This gets the project workforce to overcome spending time on low-value work and shift to the more strategic work better perceived by the executive teams. The speed of decisions could be improved with the expanded and consistent use of GenAI. In addition, the quality of decisions and directing those decisions to value becomes a core responsibility of the project manager.

This topic of decision-making also rolls up to the boardroom. Future organizations will be able to have an improved level of governance excellence in the boardroom because of responsible use of AI. As the world becomes more projectized, board directors will be more amicable to recognizing that AI is a powerful enabler to governing. Value-based decision-making will enable the minimization of decisions based on political agendas or that are done with a siloed view of the organization.

Each of the six factors reviewed above affects the potential of AI on their own in supporting workforce planning and development. Collectively, when all six factors are properly used to break down the multiple barriers that affect the scaling of AI, the project workforce could harvest multipliers of the potential of AI. It is a great area of impact that project managers in the future should keep on their focus radar. Smart use of technology to empower the future project workforce will enable better planning and higher achievement of projects' value.

In the recent 2024 survey conducted by the Project Management Institute, Generative AI in Project Management Survey, and as summarized by Figure 7.10, the survey highlights two important dimensions that aid in the understanding of GenAI and its impact on the project workforce of the future.

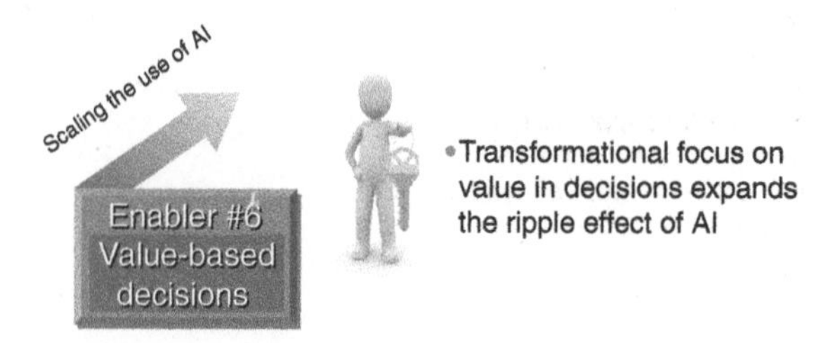

Figure 7.9 The Decision-Making Enabler

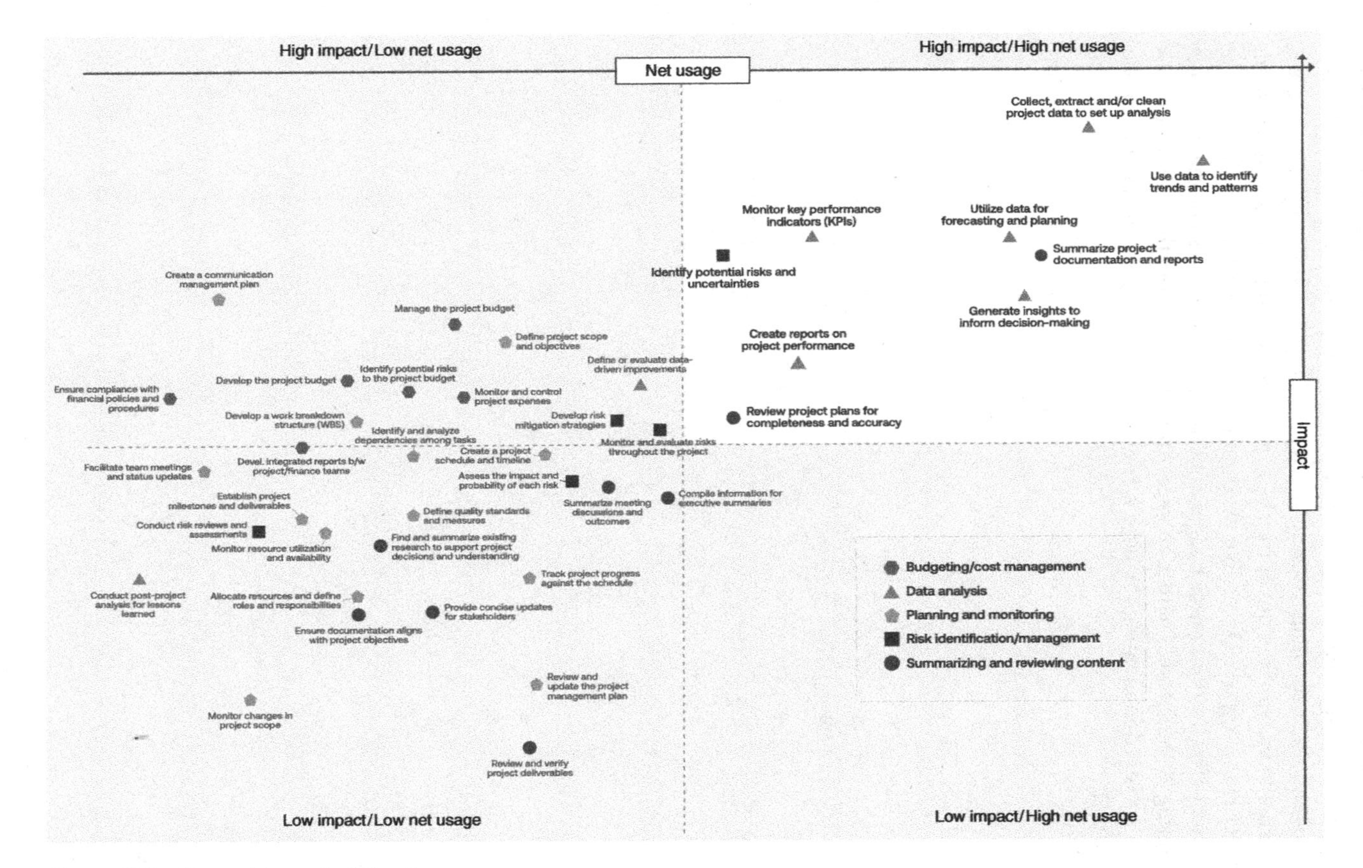

Figure 7.10 Use and Impact of GenAI

The two sides in the figure reflect net usage and impact. Especially related to the project estimating focus of this work and when reviewing the budgeting for project workforce. Cost management and planning and monitoring data, there is potential to increase the use of GenAI, especially around the quadrant where the impact is the highest. Areas like developing the project budget and identifying potential risks to the project budget are critical to the success of the workforce planning in the future. There is an opportunity to learn from the wide set of these areas surveyed in opening the door further to the likely positive use and impact of GenAI.

8

Case Studies in Workforce Planning

LEARNING OBJECTIVES

- Understanding the importance of working horizontally
- Developing a joint view of project success
- Considering additional estimate time to address environmental factors
- Driving courageous and critical project conversations
- Understanding how implementing skunk works, could enhance the agility of future ways of working

Keywords *Stakeholders' engagement; Authentic sponsorship; Understand business acumen; Risk handling strategies; Prompt engineering; Artificial intelligence and decision-making; Educating management and decision-makers*

The project workforce of the future will be under major pressure to deliver projects faster and with the highest level possible of resources' efficiencies. Pressures from shareholders and stakeholders are rising. The world of work is changing, and more projects and programs are becoming the norm of how organizations think and conduct their business. It is now the time when the percentages of projects being delivered on time, within budget, and meeting the requirements and quality will have to stop making the statistics headlines. A new era of defining success differently has been ushered in. It is expected that tomorrow's workforce will be able to avoid many of the challenges highlighted by the following case studies.

Project Workforce Estimating: Best Practices for Project Managers, First Edition.
Harold Kerzner and Al Zeitoun.

Companion website: www.wiley.com/go/Kerzner_ProjWFE

The future of work will see an extended use of iterative planning, a higher level of adaptability, and a greater risk appetite to explore. This puts pressure on project work estimating. Management will still expect the work to be free from budget variances and any contractual penalties and to have an easier time deciding whether to put on contract work. As the world of work changes, technology will augment some of the needs for higher efficiency, yet the project workforce will have to be creative, fail and learn fast, and shift their focus to strategic value achievement.

CASE STUDY: DIXON AEROSPACE

Dixon Corporation was an aerospace and defense contractor that was experiencing favorable growing pains. Dixon was winning a large percentage of the contracts they were bidding on and successfully completing them. Customer and stakeholder satisfaction was increasing.

Project managers provided the technical requirements to the functional managers, who then assigned workers they believed were best suited to fulfill the contractual requirements. No workforce planning models existed, and project managers relied heavily upon the functional managers. Project managers had virtually no input into workforce planning activities.

Unfortunately, Dixon was having a difficult time engaging team members in the projects. There were significant behavioral issues. Almost all of the project managers were engineers with advanced degrees in a technical discipline. Most of the team members on the projects were also engineers. Most of the engineers had never been trained in interpersonal skills, especially related to working on project teams and collaborating with other team members.

The organizational structure was a weak matrix organizational form where part of the daily technical direction and supervision of the team members was provided by the functional managers rather than the project managers. In addition, project managers provided no input into team member performance reviews, and almost all of the team members were working on more than one project at the same time. PMs therefore relied upon functional managers to engage the workers in the project assignments.

The decision was made by senior management that large project teams should have an assistant project manager (APM) for interpersonal skills support and human relations problem solving. The APM did not necessarily understand the technology on the projects but would assist the project manager in conflict resolution practices, reducing worker stress and potential burnout, and improving communication between team members.

Questions

1) Does this idea for an APM for human behavior issues seem like a good idea?
2) What would be a possible better solution to Dixon's behavioral problems?
3) Assuming Dixon decides to use a workforce planning model for projects, should the responsibility rest with the project manager or the APM for human behavior?

Reflection

- Consistent Stakeholders Engagement directly contributes to proper workforce planning and estimating
- Strengthening the horizontal ways of workings is a critical future competency
- Power skills (people skills) are increasingly a determinant of project success
- Although this is an aerospace and defense case, similar circumstances are visible across industries

CASE STUDY: THE PHOENIX PROJECT

Scott had heard rumors about how much trouble the Phoenix Project was in. People were leaving the Phoenix Project like rats deserting a sinking ship. Scott was grateful that he had nothing to do with the Phoenix Project. Unfortunately, all that was about to change.

Scott was a well-disciplined project manager. He was a Project Management Professional (PMP®) and was excellent in following the enterprise project management methodology his company had developed for managing traditional-type projects that started out with well-defined requirements. The forms, guidelines, templates, and checklists provided by the methodology made managing projects easy. Sometimes, Scott even believed that it was like a "no brainer" to function as a project manager. Everything was laid out for him, and all he had to do was to follow methodology protocol. Scott's company had a workforce management process, but it was used primarily by functional managers rather than project managers.

On a Friday afternoon, Scott was informed that the project manager on the Phoenix Project was reassigned, and Scott would be taking over the project. Furthermore, the executive made the following comments:

- "We were thinking about canceling the project, but first we want to see if you can turn the project around."
- "We know that we need a new business case with new ideas because we will never be able to achieve the business value expected from the original business case."
- "You'll have to come up with a different solution, one that hopefully will work."
- "I'm not sure how much support the executives can or will give you. You are out there all by yourself, except of course for your team."
- "I expect you to have a recovery plan to show me in less than a week."

The projects that Scott was used to managing started out with well-defined expectations, a clear business case, and a good statement of work. Time was no longer a luxury or even a simple constraint; it was now a critical constraint. The forms, guidelines, templates, and checklists in the enterprise project management methodology did not account for this type of situation. Furthermore, Scott had never been in any meetings that required brainstorming, problem-solving, and critical decision-making as he believed would be needed on the Phoenix Project. To make the situation worse, Scott had never worked with many of the team members assigned to the Phoenix Project. It was now quite apparent that Scott may not be qualified to function as a recovery project manager.

Questions

1) What challenges will Scott be facing?
2) What should Scott do first?
3) How important is it for Scott to familiarize himself with the current workforce?
4) Should Scott consider performing a workforce management assessment?
5) Should Scott consider replacing existing team members with new team members based upon the results of the workforce management assessment?

Reflection

- Project management success is not just about a strong methodology and great templates
- Authentic executive sponsorship is critical for proper workforce planning
- Project managers should have a voice in driving project staffing decisions
- Establishing a clear joint view of success upfront is a must have component for turning around any project across industries

CASE STUDY: BRENDA'S DILEMMA

Brenda had never been placed in such a position before while managing projects. In the past, Brenda believed that she was expected to make project-based decisions predicated upon technology. But on this project, Brenda was expected to make both business-related and project-related decisions. Previously, Brenda relied heavily upon her project sponsor for business-related decisions. But now, it appeared that most of the decisions rested upon her shoulders and the team members.

The sponsors assigned to her projects no longer wanted to hear about problems without additional data supporting some type of solution. In the past, project managers had the tendency to send their problems to their sponsors, and eventually the sponsors would resolve the problems. Soon, sponsors were spending more time solving problems on projects rather than performing the activities dictated by their job descriptions. To resolve this problem, the sponsors were now telling the project managers, "Do not come to us with problems unless you also bring alternatives and recommendations!"

Brenda was experienced in making project-based decisions where alternatives were developed based upon the constraints of time, cost, scope, quality, and sometimes risk. But making business decisions would require consideration of additional constraints, namely safety, image, reputation, goodwill, stakeholder relations management, culture, future business, and customer satisfaction.

Brenda's project team was composed mainly of engineers, many of whom had never taken any courses in business. They understood technology and how to develop technical alternatives. They knew very little about marketing and sales activities. Many of the engineers on her team were also prima donnas who believed that their opinion was the only opinion that counted.

Putting all these people in a room and asking them to develop and evaluate alternatives to the problems would certainly be difficult. Brenda did not know where to begin.

Questions

1) Given Brenda's dilemma, what are her choices for obtaining the business information she needs?
2) Which of her choices appears to be her best option?
3) Will a workforce planning assessment show if perspective team members are capable of making business as well as technical decisions?

Reflection

- Business acumen is becoming an increasingly critical competency for the future workforce
- In designing the proper workforce mix, it is important to consider the mix of complementary skills that the project team brings
- Estimating required project efforts should allow for the additional time required to investigate potential environmental factors changes

CASE STUDY: THE BRAINSTORMING MEETING

Paul was delighted that all the subcontractors were willing to send technical representatives to the brainstorming meeting. Paul's company won a contract from one of their most important clients to develop a new product. The contract involved state-of-the-art technology that was unavailable to Paul's company except through subcontracts. The client worked with Paul in the selection process of the subcontractors.

The client knew right from the start of the project that there were risks in the project and that the product may not be able to be developed without significant tradeoffs. The client's original statement of work was more of a "wish list" of deliverables with little chance of being accomplished.

Once the project started and the problems began to mount, Paul and the client jointly agreed that the direction of the project must change to salvage as much value as possible. The subcontractors had to participate in the brainstorming meeting because they possessed the expertise regarding what could and could not be done. Paul had great expectations that everyone in attendance could agree on a new direction for the project.

Paul had been in other brainstorming sessions and knew that all ideas should be listed but not evaluated or criticized until sometime later, perhaps even after the session was completed. But since several of the suppliers were not geographically local, Paul opted to conduct the meeting using the Nominal Group Technique whereby each person would present their ideas and be subject to immediate evaluation and criticism.

Each of the subcontractors presented their arguments for why their approach would be best. Even though the ideas of several subcontractors could possibly be combined into a workable solution, the subcontractors refused to budge on their position. The subcontractors' adamant position made it appear that they were more interested in how much follow-on business they could get rather than what was in the best interest of the client.

Paul now had a problem. How could he get the subcontractors to work together to come up with an agreed-upon direction acceptable to all? Paul had never been in this type of situation before. Obviously, this was not what Paul had expected as an outcome of the problem-solving meeting.

Questions

1) What appears to be the root cause of the brainstorming session problem?
2) What other mistakes were made?
3) How should Paul prepare so that problems like this do not reappear in the future?

4) Can the output of a workforce planning model identify workers that are experienced in brainstorming sessions?
5) Can workforce planning models be used to evaluate the employees in a subcontractor's organization that would be assigned to your project and interfacing with your project team members?

Reflection

- Clarity of outcomes is a cornerstone for project success
- The project manager should have proper risk handling strategies when projects have a high dependency on subcontractors
- Navigating ways to motivate subcontractors to focus on the best interest of the customer's project is an important workforce skill

CASE STUDY: THE LACK OF INFORMATION

John was an experienced project manager, at least he thought that he was. His company had an enterprise project management methodology that contained forms, guidelines, templates, and checklists for just about any project. It was a one-size-fits-all approach. Unfortunately, there were no instructions for John's current dilemma.

The statement of work was reasonably clear as to what direction the project should take. Everyone knew that the technical approach was optimistic and may not be achievable to satisfy the strategic business objectives. When the project plan was prepared, the primary planning objective was "least time" so that the product could be introduced into the marketplace quickly. But since the technical approach could not work as initially thought, it was necessary to redirect the project and create a new project plan.

John did not know whether the planning objective would still be "least time," or whether it would change to "least risk" or "least cost." John also realized that he needed additional information from Engineering, Manufacturing, Marketing, and Sales to solve his problem and prepare a new plan.

John's first attempt to collect the problem-solving information was met with resistance. John recognized quickly that information was a source of power, and these functional areas were unwilling to give John the information he needed. The lack of critical information placed John in an impossible position. Engineering would not confirm what new technical approaches were possible. Manufacturing could not provide any information on manufacturing costs without knowing the engineering design. The sales personnel could not provide any information without knowing the manufacturing costs. Marketing wanted to make sure there was a market need for the new design before providing information.

Repeatedly, John went to the functional areas asking for help. The answer was always the same: "Let me think about it and I'll get back to you." Not willing to throw in the towel yet and concede failure, John went to his sponsor, the Vice President of Marketing, and explained the situation. After explaining his dilemma, the project sponsor replied, "I can get you some information from our marketing personnel, but you are on your own with the other functional areas. Our functional silos have brick walls around them, and I have no authority over any of the resources in the other functional areas. You'll have to do the best you can." John went back to his office and began contemplating his future with the company.

Questions

1) What is the critical issue in the case?
2) What is the root cause of the critical issue?
3) What can John do to correct the situation?
4) If a project must be redirected, should a workforce planning model be used again?
5) How should the project manager respond if the model shows that different resources should be assigned and that the budget might change significantly?

Reflection

- Silos could negatively contribute to the likelihood of developing good, effective project plans
- Speaking the language of different functional areas must be nurtured as a necessary competency across industries
- Working across organizational boundaries is complex and should be considered in estimating the project workforce

CASE STUDY: THE INFORMATION OVERFLOW DILEMMA

Anne was placed in charge of a project to create a new product. Although Anne was experienced in project management, many of her newly assigned team members had little experience working on projects. Anne knew she could help them once the project was under way, but the greater problems would most likely occur during problem solving and decision making when developing the project plan.

Anne personally negotiated for the critical resources she believed she needed for the management of the project. During the project's kickoff meeting, Anne went through the statement of work in detail and conducted a workforce needs analysis to determine the makeup of the functional support personnel and the expected skills. She was convinced that everyone understood what had to be done. Anne instructed everyone to reconvene in a week with information regarding their specific efforts on the project and who would be assigned from the functional support areas. The information Anne needed from everyone was:

- Hours needed for each work package
- Grade level of the workers needed for each work package
- The cost of their efforts for each work package
- The time duration of each work package
- The anticipated risk of each work package

When the team reconvened, most of the team members came with alternatives. Many of the alternatives came with optimistic, most likely, and pessimistic estimates. Some of the team members came with historical data from five or six previous projects that were completed successfully.

This is not what Anne expected. She was now overwhelmed with information. To make matters worse, Anne knew her limitations and believed that she could not make a reasonable decision based upon the massive documentation that the team members were providing. Anne had to find a way to limit the information overflow.

Questions

1) Should Anne have expected this amount of information?
2) Is it better to have more or less information available?
3) How should Anne proceed?

Reflection

- Project workforce estimating is a complex task
- Estimating highly depends on the type and quality of data, and too much of it could be overwhelming
- Data is where the true value of GenAI and proper use of prompt engineering could open the door for project managers to better design their workforce mix

CASE STUDY: THE IMPACT OF ASSUMPTIONS

Karl was a highly talented engineer whose experience was restricted to engineering project management activities. Karl understood the constraints on the project he was now managing but was never provided with any assumptions, especially assumptions related to business decisions on projects.

For the projects that Karl had managed in the past, Karl made primarily project-based decisions related to technology. All decisions related to the business side of the project were made by the project sponsor. But for the project that Karl was now managing, Karl was expected to make both project and business decisions. The business decisions required an understanding of the assumptions.

Karl realized that the technical approach selected and the expected technical breakthrough may not be achievable. Selecting a new technical approach would certainly elongate the project and increase the costs. But changing the direction of the project would certainly have an impact on marketing and sales activities, especially on a long-term project. Some of the assumptions that were likely to change over the duration of a project, especially on a long-term project, might include:

- The cost of borrowing money and financing the project will remain fixed
- The procurement costs will not increase
- The breakthrough in technology will take place as scheduled
- The resources with the necessary skills will be available when needed
- The marketplace will readily accept the product
- Our competitors will not catch up to us
- The risks are low and can be easily mitigated
- The political environment will not change

Karl looked at these assumptions and wondered how these would impact problem solving and decision-making.

Questions

1) What is the root cause of the problem?
2) How should Karl proceed?
3) Would a workforce planning assessment have helped?
4) Should a workforce planning assessment be performed each time assumptions change significantly? If so, what is meant by "significantly?"

Reflection

- Assumptions could make or break a project
- The next generation project managers will increasingly come from different backgrounds and not purely the classic engineering route, even in industries that have been accustomed to making that workforce assumption
- A continued critical executive sponsor role remains necessary for handling strategic business decisions

CASE STUDY: NORA'S DILEMMA

Nora was now having second thoughts about whether she made the right career choice wanting a future as a project manager. She was also unsure if she should continue working for Dexter Aerospace Corporation.

Six months ago, Nora completed her master's degree in business administration with a minor in project management. She wanted a career in project management. She was hired into Dexter Corporation as a project manager after graduation. Dexter had numerous government contracts for exploratory satellites, and these were the types of projects that intrigued Nora.

Nora's first assignment was to work with Dexter's sales team to respond to a government request for proposal (RFP) for a new series of satellites to explore the properties of the sun. The RFP identified the requirements and technical specifications that had to be met for the development and testing of three prototypes. A follow-on contract would then be awarded to the winner of the RFP for the manufacturing of several satellites.

A meeting was held with the lead salesperson, Nora, and the engineering and manufacturing personnel that would be preparing the technical portions of the proposal.

The lead salesperson then made the following comments:

> "Senior management considers this potential contract as very important for Dexter's future and we must submit a winning bid. When estimating the work needed by your organizations and the accompanying costs, base your estimates upon the absolute minimum work Dexter must do to satisfy the requirements. We want to submit the lowest cost bid."

> "Then look for loopholes and omissions in the requirements stated in the RFP and prepare a list of all the scope changes and accompanying costs we could possibly generate after contract go-ahead. There are always things that the government did not consider, and must be accomplished, but let's not tell them about these things they neglected to consider other than through the scope changes we can generate after contract award."

> "Also, Dexter has never tested any products in the temperature range identified in the technical specifications. To get a step up on our competitors, let's include some wording asserting that we have done a little bit of testing in the temperature range requested and the results were promising. This should help us win the contract."

Nora could not believe what she had heard. The lead salesperson's comments seemed to violate what Nora learned in college about business morality and ethics and seemed to contradict the Project Management Institute's (PMI)'s code of professional responsibility. After the meeting was adjourned, Nora met with the lead salesperson and asked:

"Why aren't we 100 % honest with the government about all of the work that needs to be done to achieve project success and fulfill the requirements?"

The salesperson responded:

"The goal is to win the initial contract at all costs. It may look like we are 'intentionally' lying to the customer, but we simply consider this as our initial interpretation of the requirements."

"Sometimes, we might even bid the initial contract at a significant loss just to win it. Then, we push through the very profitable scope changes that most often generate significantly more profit than the initial contract. This is a 'way of life' in our industry, and you'll need to get used to it. It's probably the scope changes that will be paying your salary rather than the initial contract."

Nora then asked:

"Doesn't the government know this is happening?"

The salesperson then replied:

"Yes, I am sure they know this is happening. Once the initial contract is awarded, the government as well as other customers we have would rather go along with the approval and funding of many scope changes than to repeat the acquisition process and go out for competitive bidding again looking for new suppliers. On some of our contracts, the people that approve the initial contract and follow-on scope changes are military personnel that have just a two- or three-year tour of duty in this assignment and then get transferred to another assignment elsewhere. Whoever replaces them may then have to go before Congress or other approval agencies and explain the reasons for the cost overruns. This is how many of the aerospace and defense industry firms operate. Cost overruns are a way of life."

Nora looked at the salesperson and then said:

> "I have one more question. If you know Dexter has never done any experimentation in the temperature range requested by the client, why should we lie to the customer?"

The salesperson replied:

> "We are not lying. We consider this as just misinterpretation of the facts, or just an error in wording someone made. I am sure that somewhere in the labs are test results that we could 'recreate' confirming our wording in the proposal."

Nora read over the entire proposal prior to submittal to the government. As expected, included in the proposal was sort of vague wording that Dexter had some previous experience in testing products in the customer's temperature specification range. The proposal was sent off to the customer, and Dexter expected to hear whether they won the contract within 30 days.

In less than two weeks, Nora was asked to attend an emergency meeting with the lead salesperson on the proposal. The salesperson looked at Nora and said:

> "The government wants to visit our company quickly as see the test results we stated in our proposal about the testing we did within the specification's temperature range. How do you think we should handle their request since it may have a serious impact on who is awarded the contract? Think about your answer and let me know tomorrow."

Nora now had second thoughts about whether she should be a project manager at Dexter or anywhere else. Could this happen elsewhere, she thought?

Questions

1) Is this a common practice in aerospace and defense contractors?
2) Do people often get thrown into these types of situations?
3) Should Nora lie to the customer?
4) What should Nora do next?
5) Would a workforce management assessment have helped Nora?
6) How should Nora respond if the assessment comes up with a cost that is significantly greater than Dixon's bid?

7) Should Nora accept the lowest salaried workers to keep the cost down knowing full well that this might be contrary to the output of the workforce assessment information?
8) Is this a violation of PMI's Code of Conduct and Professional Responsibility?

Reflection

- Although this is a case for the aerospace industry, the highlighted difficult ethics situation could be seen across industries
- Project managers have a duty to be courageous and to ensure that critical conversations take place in support of proper organizational values
- This is also another case where strong, authentic leadership, is needed in the form of an executive sponsor or other governance leaders driving excellence

CASE STUDY: MANAGING RESOURCES IN GOVERNMENT AGENCIES

A government agency spent years downplaying the need for effective project management practices. Most of their projects were outsourced to contractors, and the agency assumed just a project monitoring position. The agency expected that the contractors would assign some of their best people to the agency's contracts. The agency struggled with understanding whether the contractors' costs and schedules were realistic but simply awarded the contracts to the lowest bidders.

As the costs of the contracts kept rising, the agency decided that it would be cost-effective to manage several of the projects themselves. The need for effective project management practices became evident after recognizing several project management issues resulting from project staffing practices.

The agency contracted with a project management consultant/trainer to help improve an understanding of project management and implement the necessary changes. The agency provided the consultant with the following situations that created issues.

Job Descriptions

The agency did not have formalized government service (GS) job descriptions for project managers. The role of the project manager did not fit into any of the traditional templates that the agency had for government job descriptions. As such, the agency appointed mid-level managers, usually GS-13 and GS-14 personnel, to function as project managers.

The appointment as a project manager initially was almost always a part-time position that they had to fulfill in addition to their functional role. The decisions the newly appointed project managers made always favored their functional organization rather than the project because they viewed a greater chance for advancement through the functional organization rather than through project management.

Full-Time Assignments

The agency recognized that some projects needed full-time project managers. Individuals were selected from mainly the GS-11 and GS-12 positions to serve "temporarily" as a full-time project manager and return to their previous functional department at the completion of the project. Some of these newly appointed project managers enjoyed the power and authority that came with the new assignment and wanted their projects to last as long as possible, resulting in poor project decision-making.

Co-located Teams

The agency initially supported the concept of matrix management, where project team members could be assigned to several projects on a part-time basis and remain in their functional departments. Unfortunately, there were several project managers began enjoying the power of being a project manager and disliked the idea of having to share resources with other projects. These project managers preferred the concept of co-located teams.

One project manager was assigned to a project that could have been completed in about one year with part-time team members. Instead, he created a schedule for a two-year project with all team members assigned full-time. He found a vacant floor in a government building and demanded that the entire team be removed from their functional organizations and relocated to the building where the project was now located. Senior management reluctantly agreed to his request.

The project manager did not have any wage and salary administration responsibilities even though the workers were co-located and assigned full-time. By removing the team members from their functional organizations, most of the functional managers lost contact with the team members and considered them as "non-promotable" for the duration of the project. The situation became even worse as the project came to an end. Many of the functional managers replaced the workers that were assigned to this project with other employees, and there was no longer a "home" organization for them to return to. They had to find employment elsewhere or accept reassignment to another organization that could result in having to relocate their families.

Questions

1) If you were the trainer/consultant, who would you want to train first?
2) Would the training include the effectiveness of project sponsorship?
3) How would you handle discussions about which is better: co-located teams or matrix management practices?
4) Should workforce management planning be included in the training?
5) If so, who should be trained?
6) What advice would you give management on how to determine if the workforce staffing plan for a project is correct?

Reflection

- Although the discipline of project management has been maturing over the years, there remain fundamental workforce planning issues pertaining to the matrixed nature of project work that have yet to be resolved
- The projectized nature of work that is dominating industries will require a new generation of project managers capable of directly influencing the career choices and success of their project teams
- Educating management and decision-makers on what good looks like for project management staffing needs is a critical strategic priority for organizations of the future

CASE STUDY: KANE CORPORATION

During the past several years, Kane Corporation became quite interested in the advances taking place in artificial intelligence (AI). The potential benefits that many companies envisioned resulting from the continuation of AI research were significant. Kane understood quite well that new AI inventions or technical breakthroughs would have both advantages and disadvantages. Kane's technical and project management community believed that the advantages of using AI greatly outweigh the disadvantages.

Although the rewards from successful AI implementation appeared to be well worth the effort, there were challenges that had to be considered before total company usage because some of the risks could result in significant damage or disruption to Kane. The decision was made to implement AI practices slowly, beginning with project management usage.

Costly Implementation

One of the biggest challenges facing Kane would be the AI implementation cost. IT experts at Kane estimated that the implementation cost in some organizations could conceivably be between $4 and $5 million if they had to create the software themselves. Kane believed it would be too costly to develop their own AI software and preferred to purchase a third-party package for about $2 million. AI would be used as part of workforce planning, and the concern was how to bill back to the projects some of the AI cost.

A top concern for Kane's senior management was the risk with purchased packages with how to protect themselves from outsiders that may find ways to tamper with Kane's systems and extract proprietary information. The fear was that the AI system would contain proprietary information about the capabilities of workforce employees, and if the competitors had this information, they could try to get the employees to leave Kane Corporation.

Kane believed that they could develop their own AI system if they could hire qualified resources. Universities were offering degrees in AI, and there were professional societies and companies offering certification programs in AI. Each year, more and more people were developing skills in AI applications, and this was encouraging organizations to hire these people and develop their own AI programs specifically customized for project management applications. However, there could still exist employee misuse.

Policies and Procedures

Kane decided to incur the cost of purchasing an outside package. Senior management considered what AI policies and procedures need to be put in place. This included who would be allowed to use the AI system, how partners and stakeholders would be allowed to interact with AI, who would be responsible for compliance with government regulations and laws on the use of AI, and how continuous improvements would be managed. Project management policies and procedures would need to be developed enforcing workers to use AI in an ethical manner.

Legal Responsibility

The policies and procedures that Kane wanted to develop had to be based upon legal factors, at least initially. Kane needed to see if legislation existed for the right to data privacy on certain information. Information stored in Kane's AI system would be company proprietary data on project performance or personal information that some people would not like to have shared by AI.

Stakeholder Buy-In

Another serious challenge, with possible long-term cost implications, is how Kane's project stakeholders might react to the use of AI systems on their projects. Assessing clients for use of AI as part of project monitoring, control, and status reporting may seem practical. The cost of AI could be billed on a project-by-project basis or included in the overhead structure of the firm. However, not all clients may want or agree to having AI make decisions on their projects and may be reluctant to pay for AI support.

Need for Continuous Improvement

AI systems are subject to degradation, and the software can become outdated. System support will be needed on a continuous basis. New information must be continuously added into the system and outdated material removed.

AI Data Bias

There exists a belief that, since AI is part of a computer system, it is unbiased. This is certainly not true. Data bias occurs when the data included in the knowledge repository that supports AI is incomplete or skewed. The bias may be intentional or unintentional.

Human Interactivity

Human interactivity deals with the way that workers will interact with the information provided by AI. Workers at Kane may be told to perform work differently than they are accustomed to when using AI. Workers may be told to act in a manner that subjects them to possible physical harm. This is a challenge Kane did not consider initially.

Lack of Emotions and Creativity

Kane recognized that there are applications for AI on almost all projects. However, perhaps the greatest weakness is that AI lacks the ability to introduce creativity into many project decisions. Kane believed that this may change in the future. AI may be able to provide some creative input into the implementation of an existing solution but may not be able to create totally new or innovative solutions. The need for active participation by people for creativity requirements will still exist.

Conclusions

After reading all the risks identified in this case study, you should ask yourself, "Is AI worth consideration for project management applications?" The rewards resulting from AI applications in project management certainly exceed the risks. The world appears enamored and highly interested in AI applications. We cannot predict at present the impact that AI will have on project management, but the current belief is that AI has massive potential advantages in both the short term and the long term. But as mentioned throughout this case, plans must be made to overcome the challenges.

Questions

1) What should Kane do before purchasing a third-party AI package?
2) Given that AI would be used first in project management, who should take the lead in creating the necessary policies and procedures?
3) What might be some of the issues if the AI system is used incorrectly and creates the wrong deliverables, damaged deliverables, or someone gets hurt using the deliverables?
4) What are some of the challenges Kane will face if they must work with companies that both agree to using AI and those companies that will not agree to its use?
5) What are the types of continuous improvement that will be needed and who will most likely take the lead in the continuous improvement efforts?
6) What is meant by AI data bias?

7) What is meant by human interactivity with an AI system and what bad results may occur?
8) Can AI be used effectively as part of enterprise workforce planning?
9) Can AI effectively replace enterprise workforce planning in the future?
10) Should the use of AI be billed back to each project?

Reflection

- AI is here to stay, and the possible upside of its use in project management is growing
- When estimating future project workforce, care should be placed on the extent to which AI is used in the decision-making, and the potential data bias impact on the workforce's way of working that AI might introduce
- The responsible use of AI should be a strategic organizational choice

CASE STUDY: SKUNK WORKS PROJECT MANAGEMENT

Today, regardless of what periodicals or books you read that discuss project management practices, you will most assuredly find information discussing Agile and Scrum. What most people do not realize is that several of the principles of Agile and Scrum are more than 80 years old, having been used by Lockheed during the 1940s when it created the famous "Skunk Works" dedicated to radical innovation. Most people may have heard of "Skunk Works," but do not understand the impact it had on project management practices years ago and the impact it is still having in many companies worldwide.

The Need for an Innovation Unit

One of the main drivers of a company's competitive advantage is innovation. Unfortunately, there are several types of innovation, and each type comes with advantages and disadvantages that may affect certain functional units. Let's consider just incremental (or a continuous small improvement) innovation and radical innovation.

The selection of the type of innovation can be impacted by the personal desires of the people that must make the decision and is often based upon how they feel about the status quo versus the future. Some companies are fearful of the radical innovations from Skunk Works because of the risks in accepting new businesses. Examples would include Xerox and personal computers and Kodak's failure in digital photography. These companies focused mainly on the expansion of core businesses.

Many executives prefer to promote short-term results, such as in established businesses that generate sales, profits, and current executive compensation and reward packages, rather than radical innovation where the results may not be known for years and are accompanied by financial uncertainties. When executives resist major changes, they then assign their brightest and most talented people to short-term results and commercialization of new ideas may suffer.

Functional units can also resist new technologies if there is a fear of being removed from their comfort zones. Changes in technology can be accompanied by additional costs such as purchasing new equipment and facilities, hiring new workers, developing of new procedures, retraining expenses, and new marketing and sales requirements.

The resistance to changes in technology can trigger competition between functional departments such as R&D and manufacturing. The unfortunate result in some companies is when manufacturing resists radical innovation practices that could favorably impact the organization's future. To overcome the resistance problem, companies created a so-called Skunk Works unit for radical innovation where the unit is isolated from the parent organization. The traditional R&D organization would then be responsible for continuous improvement projects, and the Skunks Works unit would manage radical innovation activities.

The Birth of "Skunk Works"

During the early years of World War II, the United States and its allies realized that their fighter planes were no match for Germany's new jet fighters. The U.S. War Department asked Lockheed for help in 1943. Lockheed created a special unit entitled Lockheed's Advanced Development Program. Later the name of the program was given the pseudonym "Skunk Works."

The term "Skunk Works" came from Al Capp's hillbilly comic strip Li'l Abner, which was popular in the 1940s and 1950s. The original term in the comic strip, "Skunk Works," was a dilapidated factory that generated strange odors and was located on the remote outskirts of Dogpatch. The Lockheed unit began using the term "Skunk Works" thanks to an engineer in the original team that was a fan of the comic strip.

The special unit was headed up by Kelly Johnson, Lockheed's 33-year-old chief engineer, who ran the unit for almost 45 years. His nickname at Lockheed was "Engineer of the Century." The intent was to create a special team composed of a small group of some of Lockheed's most talented employees, handpicked by Kelly, to work on secret projects that required innovation. To help maintain secrecy and avoid distractions, the team was allowed to work autonomously at a secret location away from distractions that could come from Lockheed's main operations. Kelly was provided with a limited budget to support the effort and aggressive schedules.

From a project management perspective, Kelly was the program manager responsible for all of the secret projects within Skunk Works. However, as chief engineer at Lockheed, he also had to share his time each day with ongoing activities at the main operations unit that were not part of Skunk Works. Kelly was highly successful in his tenure of running Skunk Works for 45 years.

During his tenure, Kelly developed 14 "Rules" for all Skunk Works projects that were directly related to most project and program management practices requiring innovation. The "Rules," most of which still apply today, will be discussed later in this section. Ben Rich, who eventually replaced Kelly, also promoted the 14 "Rules." The result has been an ongoing record of innovations at Lockheed for more than 70 years.

Challenges with "Skunk Works" Growth

Companies that need innovation for growth and survival have heard of Skunk Works and recognize the application for running secret projects using the best people available. Skunk Works thrive on self-driven teams that focus on making breakthrough innovations in a reasonably short time frame. The selection of the researchers for Skunk Works is critical. They must enjoy the research and experimentation needed in dealing with the risks and uncertainties that could lead to major innovation breakthroughs. They must also possess a passion for teamwork and cooperation with colleagues at Skunk Works. A knowledge of project management is most certainly helpful.

Skunk Works shows the entire company where technology may be heading, and this is accomplished without spending a great deal of money. The result is most often better decision-making on opportunities involving creativity. As stated by May Matthew,[1]

> "High-quality designs in a short time frame with limited resources are the hallmarks of a Skunk Works project."

The Skunk Works approach has been used successfully by numerous companies. Steve Jobs used it to launch the Macintosh computer at Apple as well as the iPhone and iPad. Ford Motor Company used Skunk Works to rapidly integrate technology into useful automotive features. Disney created an entire division entitled "Imagineering" (i.e. IMAGination and engINEERING) to function as an R&D laboratory to bring stories to life. The division is remotely located from Disney's headquarters and functions as the creative unit that designs and builds all Disney theme parks, resorts, cruise ships, games, publishing, movies and cartoons for TV and theater screens, and product development businesses. IBM used Skunk Works to create personal computers. Microsoft also used Skunk Works to develop computers and tablets. HP created pocket calculators, laser printers, and 3-D printers using Skunk Works.

---

1 Matthew, E.M. (2013). Skunk works: how breaking away fuels breakthroughs, *Rotman Management*. Spring, 52–56.

Other well-known companies using Skunk Works included Google, DuPont, Boeing, GenCorp, Siemens, Philips, Intel, LEGO, and Xerox. The management guru, Tom Peters, co-authored a book entitled "A Passion for Excellence" in which Skunk Works was highly praised as a means for innovation, competitiveness, and growth.

Some companies focus on part-time innovation. Google's "20% time" policy allows employees to spend one day a week working on projects even though they may have other responsibilities. The results were Google News and Gmail. To promote this policy, Google demands that at least 30% of each division's revenue come from products introduced within the past four years. This impacts employees' bonuses and salary. A similar policy exists at 3 M. Employees are allowed to spend 15 minutes each day thinking up new products for 3 M, and at least 25% of the division's revenue must come from products introduced within the past five years.

Most of today's companies have recognized the need for innovation, creativity, design thinking practices, and advances in technology. Yet many of the companies have not given consideration for Skunk Works as a possible means for growth because of their interpretation and fear of the accompanying challenges. Lockheed was able to overcome the challenges, but even with their success, they admitted there will be limitations for others. Ben Rich, who served as Vice President and General Manager at Lockheed's Skunk Works, discussed the challenges some companies will face based upon his experience with government projects[2];

> "I seriously doubt that most of these companies will successfully implement the Skunk Works' management style, however. In many, if not most, cases, it's the wrong thing to do. There are too many outside factors that hinder implementation of the Skunk Works' philosophy, not the least of which is the number of requirements imposed by the United States government."

Even today, Skunk Works is considered by many as restricted mainly to large and expensive high-technology projects specifically designed for aerospace and defense units of the U.S. Government. This is certainly not true. Lockheed's success has been with small as well as large projects requiring creativity.

---

2 Rich, B. R. The skunk works management style: it's no secret. *Vital Speeches of the Day*. 11/15/88, 55 (3), 87–93.

Results have shown that the successful marriage between Skunk Works and project management practices can lead to innovation efficiency. Unfortunately, an innovative product, even quickly, is no guarantee that there will exist a market demand for the product. There must exist a business need for creating a Skunk Works unit. Some units fail to develop a strategy for commercializing the innovation outcomes. Peter Gwynne identified challenges that Xerox faced and how they addressed the challenges[3];

> "To be successful with them (Skunk Works), they have to be business oriented – that is, they must create successful businesses rather than successful products. So, Xerox is now taking a new approach to Skunk Works: Starting up projects as small businesses with their own Profit and Loss (P&L) responsibility and marketing personnel, rather than internal groups that have to rely on the corporation for those activities and people."

In most of the Fortune 100 companies, project management is more than just another career path. It is seen as a strategic competency necessary for the growth of the organization. As a project manager, you are now seen as managing part of a business rather than just a project. You are expected to make business decisions as well as traditional project decisions. Most projects today that focus upon innovation outcomes include a life cycle phase entitled commercialization.

Unlike traditional product improvement R&D that might focus upon finding higher quality raw materials or cheaper ways to manufacture the products, Skunk Works have a significant business component that includes prototype development, reducing time-to-market, developing their own channels of distribution, and selling the products directly to the customers.

Developing innovative products does not maximize business benefits to a company unless the innovation team is allowed to make the necessary time-to-market commercialization decisions to take advantage of opportunities. As stated by Single and Spurgeon in discussion about Ford's Skunk Works,[4]

> "All automotive companies are working hard, with considerable success, to reduce the time from concept to customer for vehicles. It is necessary to do the same thing for innovative features. Companies

3 Gwynne, P. (1997). Skunk works – 1990s style. *Research Technology Management*, 40 (4), 18.
4 Single, A. W. & Spurgeon, W. M. (1996). Creating and commercializing innovation inside a skunk works. *Research Technology Management*, 08956308, 39 (1), 38–41.

that learn how to do this will certainly have a competitive edge. A well-designed Skunk Works is an eminently practical way of accelerating the implementation process."

Another challenge with Skunk Works is the culture that is created. Implementing Skunk Works has forced senior management to rethink the issues with allowing multiple cultures to exist concurrently. For years, companies allowed each project to have its own culture, knowing that the projects would eventually come to an end. As companies began realizing that project teams must make business decisions, a single corporate culture was created in many companies that supported all types of projects and traditional business practices. Skunk Works cultures in most companies appear to be business-oriented, but they must also be product-innovative-oriented. As such, most people view Skunk Works as countercultural to protect the team from possible disagreements with the corporate culture.

Cultural differences can lead to misalignment issues. Misalignment in the relationship between the primary organization and Skunk Works. The greater the misalignment, the greater the chance that some good opportunities might be discarded and other ideas might be promoted that are too risky and not in line with corporate goals and objectives.

Skunk Works thrives when team members can use unconventional approaches to problem-solving and decision-making, regardless of the size of the projects or programs. This often scares some executives who are afraid that implementation might cause senior management to lose control of the company by eliminating bureaucratic red tape needed for product checks and balances and reducing the time needed for approvals and decision-making. Skunk works have minimal managerial constraints.

Project management has matured significantly since Ben Rich delivered his speech (see footnote 2) more than 30 years ago. The benefits of using project management and the accompanying best practices appear in numerous publications. Yet there still exists inherent fear of the Skunk Works approach in some organizations. In many companies, project and program governance still resides at senior management levels because executives do not trust project teams to make certain decisions that were traditionally reserved for senior management. Senior management also preferred to monopolize customer communications. This is contrary to what Lockheed did by allowing project teams the autonomy to develop close working relationships with customers and stakeholders.

There is also the fear among companies that might have government contracts that they will be expected to allow heavy involvement by government stakeholders and be burdened with an excessive number of legal policies and procedures that must be followed. Companies may fear that this may trickle down to non-government contracts as well. This is especially true with the growth of AI applications and concern over possible product liability lawsuits. Companies may not realize that many of these outside factors that existed previously had been restricted only to government projects and programs.

Another critical issue is the size of the company. As stated by Ben Rich,[5]

> "I don't think a 'Skunk Works' would be feasible if it couldn't rely on the resources of a larger entity. It needs a pool of facilities, tools, and human beings who can be drawn upon for a particular project and then returned to the parent firm when the task is done."

Company size today is no longer an issue for successful project management to exist but may impact the decision to implement Skunk Works. Even the smallest of companies can implement successful project management practices.

Some companies have used Skunk Works to respond to a customer's RFP. The response might include a prototype that underwent inspection and testing. If the company's bid is not accepted, the unit is dissolved, and people return to their previous organization. If the bid is accepted, the unit begins commercialization.

In some extreme situations, a company might establish multiple Skunk Works to bid on the same RFP. In this situation, the units are also in competition with each other to win the opportunity to submit their bid using their designs and therefore do not communicate with each other or share information. These units may include contracted labor.

Management would select the best innovative approach from one of the units for their bid. The other units are then dissolved. There are several risks with doing this. Other than cost, the use of contracted labor can create issues with secrecy and control and ownership of intellectual property.

Perhaps the most important lesson learned from Skunk Works is the need to develop a corporate project management culture that can bring out the best in people, and this often requires an unconventional approach to project leadership where team members are effectively engaged throughout the life of the project or program. If this is done correctly, creativity will follow and lead to success. The challenges and issues can be overcome.

---

5 See footnote 2.

Kelly's 14 Rules and Practices at Skunk Works[6]:

Kelly's 14 rules were specifically designed for Lockheed's Skunk Works. Today, most of these rules still apply and may be highly beneficial to all companies, especially those needing innovation and creativity. The rules are designed around project and program management practices that have been highly successful at Lockheed for more than 70 years. The rules will be discussed from a project management perspective.

Rule #1. The Skunk Works® manager must be delegated practically complete control of his program in all aspects. He should report to a division president or higher: Innovation decisions that must follow the chain of command and obtain everyone's input and approval can be time-consuming and slow down the decision-making process. Project governance on many types of projects works best with individual rather than committee sponsorship, and that individual should reside near the top of the organizational chart. Single-person governance can also eliminate having to work with often hidden agendas of many managers that wish to participate in decisions involving innovation for personal reasons rather than for what is in the company's best interest.

Rule #2. Strong but small project offices must be provided both by the military and industry: Not all projects and programs can be managed by a single person. Some projects require the creation of a project office composed of APMs. Clients like the U.S. government often demand that a government project office also exist on site as a means of tracking performance on some high-visibility government programs. When this occurs, contractors often assign the same number of people in their project office as the customer would have in their project office to provide one-on-one coverage and communications. Large project offices increase overhead, increase communication channels, slow down decision-making, and increase the project's overhead costs.

Rule #3. The number of people having any connection with the project must be restricted in an almost vicious manner. Use a small number of good people (10–25% compared to the so-called normal systems): Strategic projects, especially those that require innovation and creativity, have a much greater need for problem-solving and decision-making practices. The larger the

6 The rules can be found at lockheedmartin.com/us/aeronautics/skunkworks/14rules.html.

number of people connected to the project, the greater the number of channels of communication that must exist. This can take a great amount of time and increase a project's budget. By restricting the number of people connected to the project, decisions can be made in hours or days rather than weeks or months. Action items are more quickly resolved.

Rule #4. A very simple drawing and drawing release system with great flexibility for making changes must be provided: This rule was created before we had computer-aided design systems, Computer Aided Design and Computer Aided Manufacturing (CAD–CAM). The intent, which still exists on projects requiring drawings, is to make it easy for changes to be made and approved.

Rule #5. There must be a minimum number of reports required, but important work must be recorded thoroughly: We have all written reports that are never completely read. Report preparation is costly and involves writing, typing, proofing, editing, approvals, reproduction, security classification if necessary, and even disposal. The cost, fully burdened for everyone involved, could exceed $2000 per page. Reports should be minimized but accurate and include all the critical information.

Rule #6. There must be a monthly cost review covering not only what has been spent and committed but also projected costs to the conclusion of the program: This rule has become standard as a part of all project monitoring and control reporting systems. Reporting today includes the estimate at completion (EAC) as well as actual and budgeted costs.

Rule #7. The contractor must be delegated and must assume more than normal responsibility to get good vendor bids for subcontract on the project. Commercial bid procedures are very often better than military ones: Even in today's environment, government customers in some countries still dictate to contractors how to evaluate suppliers and which suppliers they can use. In one country, the local government forced contractors to select suppliers only from within the country and to give favoritism to suppliers in cities that had the greatest unemployment rates. Topics such as cost, quality, and lead times were of secondary importance.

Rule #8. The inspection system as currently used by the Skunk Works, which has been approved by both the Air Force and Navy, meets the intent of existing military requirements, and should be used on new projects. Push more basic inspection responsibility back to subcontractors and vendors. Don't duplicate so much inspection: As discussed in Rule #2, customers and government agencies often establish project offices on the contractor's site. This can lead to duplication of inspection practices and can force contractors to establish multiple inspection processes based upon customer requirements.

Rule #9. The contractor must be delegated the authority to test his final product in flight. He can and must test it in the initial stages. If he doesn't, he rapidly loses his competency to design other vehicles: Allowing government and military personnel to have the responsibility for product testing can create issues if the personnel are rotated to different assignments during the project and new people appear with a different interpretation of how good the product works. Product testing is the responsibility of the company that must design and manufacture the product. Testing should be done throughout the life cycle of the project to minimize the risks of downstream product liability lawsuits.

Rule #10. The specifications applying to the hardware must be agreed to well in advance of contracting. The Skunk Works practice of having a specification section stating clearly which important military specification items will not knowingly be complied with and reasons therefore is highly recommended: If appropriate, all contracts and even statements of work should have a specification section. Project teams must clearly understand specification requirements before the final contract price is agreed to.

Rule #11. Funding a program must be timely so that the contractor doesn't have to keep running to the bank to support government projects: Customers often underfund contracts just to get the work started. Contractors often grossly underbid the initial contract and then either ask for additional funding or try to push through scope changes. In either case, both the contractor and customer must have a clear understanding of the cost of the project and the available funding to match the cost.

Rule #12. There must be mutual trust between the military project organization and the contractor the very close cooperation and liaison on a day-to-day basis. This cuts down misunderstanding and correspondence to an absolute minimum: Trust has become perhaps the most important word in project management. One of the reasons why customers establish a project office at the contractor's location is to minimize paperwork and reduce misunderstandings. Customer communication in the past was at a minimum because contractors believed that customers and stakeholders did not understand project and program management and would meddle in the daily operations of the projects. Today, customers and stakeholders possess project management knowledge, and their help and advice are welcomed.

Rule #13. Access by outsiders to the project and its personnel must be strictly controlled by appropriate security measures: Outsiders often go to extreme measures to find out what projects your company might be working on to bring this knowledge back to their organization. One company even went so far as to find out the salary of certain people working on secret innovation projects and then offered them a larger salary to change companies.

Rule #14. Because only a few people will be used in engineering and most other areas, ways must be provided to reward good performance by pay, not based on the number of personnel supervised: People should be paid and rewarded for their accomplishments rather than the size of their empire.

Project Management Practices within Skunk Works

The ability of the team to collaborate with each other, respect each other's opinion, and a willingness to participate in group decision-making are mandatory for increasing the chances of Skunk Work success. These are some of the reasons why the participants in Skunk Works are most often hand-picked by the leader. Gaining the benefits of a successful Skunk Works may require organizations to rethink how project management should be implemented within the unit. Flexibility and the use of techniques such as Agile or Scrum are beneficial. Projects that have a heavy focus on innovation often follow different practices than traditional projects that begin with well-defined requirements that may remain fixed over the life cycle of the project. Project management practices within the Skunk Works should include the following:

- Project planning may need to be structured around short time periods, such as sprints.
- At the end of each period, continuous improvement decisions must be made based upon experimentation, inspection, observation, and experience.
- Team members must recognize the need for continuous feedback and that project success is based upon iterative development.
- Teamwork should be seen as the driver for success.
- Project team members must respect each other, and the recommendations and decisions others might make.
- Collaboration with team members and stakeholders is more important than relying upon tools and processes.
- Project documentation should be minimized if possible.

- Project teams must be prepared to make business and product commercialization decisions.
- Project teams must be willing to be removed from their comfort zones and work on tough problems.
- Safe guarding intellectual property is critical.
- Business metrics that focus upon business goals and objectives should be used along with traditional project metrics.

Many of the above bullets are the characteristics of Agile and Scrum project management practices. There are certainly other factors that could be included.

Conclusion

The need for innovation and new products will most certainly increase. In the future, more companies are expected to consider Skunk Works as a possible solution to corporate growth. Combining the above bullets with Kelly's 14 Rules provides us with a glimpse of how project management practices take place in Skunk Works. Effective project management practices can lead to innovation and commercialization success. But it will be challenging for some companies.

Questions

1) Why was Skunk Works at a remote location?
2) What are the differences between traditional project workforce planning and Skunk Works workforce planning?
3) Is it a good idea for the leader of Skunk Works to personally select the workforce?
4) What was the rationale for wanting to keep the size of the workforce small?

Reflection

- Skunk Works principles have been in place for decades and are directly valuable to innovating the future of organizations
- Implementing Skunk Works affects the ways of working and exhibits many similarities with the working environment and expectations of agile teams
- In designing and estimating future project workforce involved in innovation initiatives, care should be given to the cultural implications and the ground rules necessary to support the success of Skunk Works

Index

A

absorbed slack, 82
activity metrics, 91
adjusted workforce smoothing, 123
Agile, 18, 34, 62, 83, 170
analogy estimating, 38–39
artificial intelligence (AI), 8, 18, 45, 125, 156
assistant or deputy project managers (APM), 117, 120
assumptions tracking, 29

B

backup plans, 45–46
blended rate, 50
bootlegged projects, 79
brainstorming capability, 85
brainstorming meetings, 43, 141
Brenda, 139
budgetary estimating, 38

C

career enabler, 128
close innovation, 66, 67
co-creation innovation, 69–71
co-creation partnership, 71
co-creation projects, 70
co-located teams, 120
competitive bidding, 119
complete failure, 72
complete success, 72
computer aided design and computer aided manufacturing (CAD-CAM), 168
construction organizations, 53
continuous innovation, 62
contract profitability, 116
core metrics, 90
Corporate General and Administrative (G&A), 28
cost estimating relationship (CER), 40
cost-reimbursable contracts, 55
creativity, 85
crisis-driven innovation, 68
critical success factors (CSF), 89
critical *vs.* non-critical constraints, 30
CSF *see* critical success factors (CSF)
customer communication, 169
customer-funded contractual changes, 116

Project Workforce Estimating: Best Practices for Project Managers, First Edition.
Harold Kerzner and Al Zeitoun.

Companion website: www.wiley.com/go/Kerzner_ProjWFE

D

decision-making, 7, 78
decision-making enabler, 130
degrees of permissiveness, 103–104
design thinking, 84
detailed or activity estimation, 36
Dexter Aerospace Corporation, 149
diamond of innovation, 68
direct *vs.* indirect project costs, 23
discontinuous innovation, 68
disruptive innovation, 65, 68

E

earned value measurement systems (EVMS), 21, 22, 50
earned value measurement techniques, 89
end-to-end project management, 63
enterprise risk management (ERM), 47–48
estimate at completion (EAC), 168
experience curves, 40

F

failures, 72
fast-changing organizations, 81
FFE *see* fuzzy front end (FFE)
financial metrics, 92
followership innovation, 68
follow-on contract, 149
forward pricing rates, 25–26
full-time employees, 10
fully burdened costs, 23
fully loaded costs, 23
functional employees, 105
functional managers, 36, 105
functional organizations, 45, 119
future workforce
  collaboration, 101
  commitment and expectation management, 104
  degrees of permissiveness, 103–104
  high-value team members, 105
  non-financial incentives, for motivation, 106
  non-monetary rewards, 107–109
  organizational development, 110
  plaques to public acknowledgments, 106–107
  project management, 102–103
  public recognition, role of, 107
  staffing, timing and patterns of, 109
  team competencies, 101–102
  underperformance, 105
  workforce recruitment strategies, 103
fuzzy front end (FFE), 30–32

G

gap analysis, 4
generative AI (GenAI), 125, 128, 131, 132
government service (GS), 153
ground-up (grassroots) estimation, 39

H

head-count metric, 53
hidden innovation, 68
hidden workforce costs, 44
high-value team members, 105
human-digital bridge, 125
human leadership, 13
hyperbolic functions, 40

I

imagineering, 86, 162
impact metrics, 91
incremental innovation, 64, 65
indirect costs, 23, 28, 42
innovation
  agile power, 83–84
  business value extraction, 93–95
  co-creation innovation, 69–71
  and creativity, 61–62
  culture, 77–78
  environment, 74–77

foundation, 63
for growth, 62
hybrid approaches, 83–84
idea generation, 63, 78–79
management, 64
measurement techniques, 90
metrics, 89–93
overview, 62–64
portfolio management, 87–89
project management future skills, 84–86
prompt engineering, in fostering innovation, 95–96
resources management, 80–82
reward systems, 79–80
success and failure, 72–73
types, 64–69
value, 74
innovation portfolio project management office (IPPMO), 87, 88
integrated human-digital future workforce, 118
inventory skills matrix, 111

K

Kane Corporation, 156–158
key intangible performance indicators (KIPI), 90
knowledge management, 85, 86

L

labor, 36, 40–43
labor rate structures, 9–10
leadership, 12
learning curve effect, 41–42
learning curves, 40–41
least cost, 143
least risk, 143
least time, 143
life cycle costing approaches, 121
local currency, 117
long-term manpower planning, 6
long-term workforce planning, 4

M

management reserve, 116
management support cost, 42, 43
management support rules of thumb, 43
manpower, 28
planning, 6
smoothing, 123
maturing competencies, 124
medium-term manpower planning, 6
metrics, 30, 33, 91
milestone planning estimation, 35
multiple project portfolio management (PPM) tools, 34

N

natural language, 95
Nominal Group Technique, 141
non-cash awards, 106
non-financial incentives, 106
non-monetary rewards, 107–109
non-project-driven companies/ non-project-driven organizations, 2, 57, 102
Nora, 149–151

O

one-size-fits-all approach, 143
open innovation, 66–67
organizational dependencies, 112
organizational design, 59
organizational development, 11, 110, 112, 113
organizational or project manpower strategies, 4
organizational slack, 82
outcome-based reward system, 79–80
overhead costs, 24–25

P

paperless project management, 44
parametric estimating, 38
partial failure, 72
partial success, 72
permanent worker, 10
Phoenix Project, 137
PMBOK® Guide, 22, 63
poor manpower estimating techniques, 46
potential slack, 82
power skills, 127
PPM *see* project portfolio management (PPM)
predecessor models, 12
preliminary estimation, 35
pricing out projects, 116
process-based reward system, 79
procurement, 26, 115, 117, 147
product commercialization, 63
professional development, 9
project-driven companies/project-driven organizations, 2, 57, 102, 116
project management, 63–64
Project Management Institute (PMI), 62, 150
project management office (PMO), 72
Project Management Professional (PMP®), 107, 137
project managers, 8, 33–34
project portfolio management (PPM), 32–33
project portfolio software tools, 33–34
project staffing, 5
project workforce planning
  AI impact, 11–15
  future leadership qualities, 13
  future model, 3
  labor rate structures, 9–10
  limited resources, 1–3
  management principles, 3–6
  professional development, for project teams, 8–9
  role of contract and temporary staff, 10–11
prompt engineering, 95–96

Q

"quick and dirty" estimates, 38

R

radical innovation, 64
ranking process, 33
request for proposal (RFP), 149
resource allocation, 82
resource identification, 81
reward systems, 79–80
RFP *see* request for proposal (RFP)
risk appetite enabler, 129
risk management, 47
Riverside Software Group (RSG), 111–113

S

salary, 25–26
scoring models, 33
Scrum, 170
S-curve, 53
secondary values, 74
short-term manpower planning, 6
Skunk Works, 160
  growth, 162–166
  history, 161–162
  innovation, 160–161
  Kelly's 14 rules, 167–170
  project management practices, 170–171
spending curve, 54
staffing plan, 109
stakeholders, 20
standard project management methodologies, 76
state-of-the-art technology, 141
strategic clarity, 112

T

team dynamics, 14
technical prima donna, 105
temporary workers, 10
termination liability, 55
traditional project management, 70, 76
traditional *vs.* innovation project management practices, 75
transformational metrics, 91
trust foundation enabler, 129

U

unabsorbed slack, 82
undistributed budget, 116

W

work breakdown structure (WBS), 21, 35
worker availability, 18
worker career development, 8
workforce agile approaches, 83
workforce estimation
- available working hours calculation, 26–27
- direct *vs.* indirect project costs, 23
- EVMS, 21, 22
- factors, 18–20
- forward pricing rates, 25–26
- methods
  - accuracy enhancement, 36–37
  - analogy estimating, 38–39
  - backup plans, 45–46
  - challenges, 46–47
  - costs per hour, 37–38
  - ERM, essential value of, 47–48
  - ground-up (grassroots) estimation, 39
  - hidden labor costs, 43–44
  - labor costs documentation impact, 44–45
  - from labor hours to labor costs, 36
  - learning curves, 40–42
  - management and support, 42–43
  - overview, 35–36
  - parametric estimating, 38
- overhead costs, 24–25
- overview, 21
- project pricing, 27–28
- sources for, 17–18
- stakeholder involvement, in workforce staffing, 20
- validating assumptions, 28–30
- value proposition, 33–34
- work authorization form, 27

workforce expenditures
- analyzing spending trends, 53–55
- analyzing workforce metrics, 52–53
- balancing hours and dollars, in project tracking, 50–51
- business model optimization, 58–60
- documenting challenges, in workforce reporting, 57–58
- managements, 55–56
- reporting intervals, 57
- termination liability, 55
- tracking, 49–50
- work hours into financial metrics conversion, 50

workforce leveling technique, 121–122
workforce management, 3
- in external environment, 6–7
- and legislation, 7–8

workforce oversight, 56
workforce peaks and valleys, 122
workforce planning
- additional project funding, 116
- advanced workforce leveling strategies, 122–123
- AI factors
  - career, 127–128
  - culture for, 126
  - human skills building factor, 126–127
  - risk appetite, 129
  - trust foundation, 128
  - value-based decision-making, 130

workforce planning (*Continued*)
budget allocation and adjustment, 115–116
case studies
assumptions impact, 147
Brenda, 139
Dixon Corporation, 135
information overflow, 145
Kane Corporation, 156–158
lack of information, 143
Nora, 149–151
Phoenix Project, 137
resource management, in government agencies, 153–154
Skunk Works, 160–171
co-located teams, 120
contingency plans, 119–120
global workforce estimation challenges, 116–117
inventory of, skills and competencies, 123–124
life cycle costing approaches, 121
management plan data, 118–119
models, 6
team structuring, 117–118
workforce leveling technique, 121–122
workforce staffing, 20
workforce tracking, 51
writing skills, 45